THE BRIGHT AND GLOWING PLACE

Other books by Frank Rowsome, Jr.

Trolley Car Treasury (1956)

They Laughed When I Sat Down (1959)

The Verse by the Side of the Road (1965)

Think Small (1970)

THE BRIGHT AND GLOWING PLACE

FRANK ROWSOME, JR.

WITH DRAWINGS BY

ALDREN A. WATSON

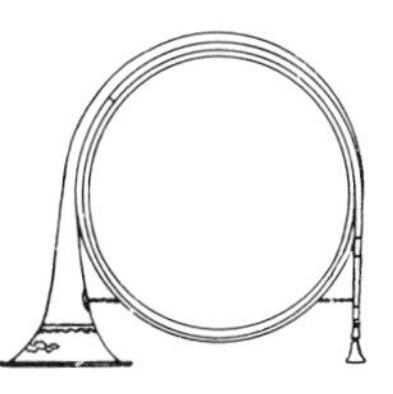

THE STEPHEN GREENE PRESS

BRATTLEBORO, VERMONT

This book has been produced in
the United States of America:
designed by R. L. Dothard Associates,
composed by American Book–Stratford Press,
printed by Halliday Lithograph Corporation,
and bound by The Colonial Press.
It is published by The Stephen Greene Press,
Brattleboro, Vermont 05301.

Library of Congress Cataloging in Publication Data
Rowsome, Frank.
The bright and glowing place.
1. Fireplaces. 2. Fire. I. Title.
TH7425.R68 644'.1 75–8195
ISBN 0–8289–0253–4

75 76 77 78 79 9 8 7 6 5 4 3 2 1

CONTENTS

THE HOME FIRES BURNING

TRUDGING back bone-cold from a walk on a winter afternoon, aware of a fire blazing at home, resolutely denying the cold ache of fingers, ears, and toes, it is easy to hallucinate the elusive fragrance of woodsmoke. This is surely among the most delightful of all smells, rich in evocation of comfort and ease. When the house is finally in sight and the blue-grey column can be discerned curving up from the chimney until lost against the slate sky, it can be a matter of only mild interest to note that the column is curved *away* from you. No matter whose fire, if anyone's, has come to your nose; the important thing now is to come inside, stomping the snow from your boots, shutting the door solidly against winter. The odor within will not be faint and elusive. It will be the characteristic smell of a house where a bright fire is blazing, an intricate compound of woodsmoke and warm paneling, with trace smells of hot hearth and ashes: a delicious *woodeny* smell. Gloves, cap, coat, and scarf are peeled off, and you come directly to the fire.

A rite in praise of the spirits of the hearth is then performed, originating in God knows what ancient prescription for human behavior. It begins formally as hands are held out to the flames, palms forward and fingers outstretched in a gesture of greeting and thanks. Then comes an interlude of improvisation in the fire-dance, a warming of ears, cheeks, and wrists, a blowing and rubbing of the nose. Men do a symmetric flexing of arms and shoulders; women go through a curious episode of embracing themselves, hands crossed to upper arms in a unique posture that mimics the classic female stance complaining of cold. In the freestyle part of the fire rite there is often a sensuous episode of unlacing boots to liberate icy feet—taking footgear off is of course one of life's most consistently reliable pleasures—kneading the arch and instep and numb toes, and then cocking the feet on the fireplace fender as sensation comes creeping back. The closing movement of the firedance comes in a few minutes when, standing in a comfortable tiredness, you turn your back to the blaze, soaking up warmth in back, rump, and thighs, absorbing heat to the edge of discomfort, staring away from the fire in a far-focused manner that could be inherited through thirty-five thousand generations from predecessors who also stood with their backs to the blaze, scanning past the mouth of the cave into the perilous darkness beyond.

The fire itself needs to have been well laid and kindled, brought past its skittish youth to an established, mannerly blaze not needing touch of poker or toe. No firescreen should be in place, and especially not that lunatic contemporary abomination, a glass screen. This means that nonsparking hardwood must be burning, husky billets of beech or oak or maple, a year or more from the axe and half of that under cover. (Unseasoned wood snivels.) An array of flames should be dancing accompaniment, pouring between the front and upper logs, each flame the color of a pale Sun, with flicks of rich blue near its base and edges intermittently tipped in red. In the ample bed of ashes beneath, a glowing line of orange-red coals will have begun to collect, promising the hypnotic shimmery-walled caves and caverns that will appear later in the fire's life, ideal for study in the relaxed, fractionally asleep state that a woodfire's radiance soon induces.

That, or something like it, is the form of a pleasing winter-afternoon or Sunday fire. There is a morning fire, very different, that at first leaps and flickers in the mind only, endowed with exceptional warmth and comfort. This is the fire imagined as you awaken, early, quite slept out, in a cold country house. It is clear from the way breath hangs in the air that no comfort whatever can be found beyond the blankets until someone is conscientious enough to get up and build a roarer.

It is a two-tempo process and the first is swift. On with the bathrobe and slippers; stride to the woodshed for an armload of wood and kindling. (Something about the fire-induced somnolence of the evening before usually prevents forehandedness.) Stride to the hearth, fling aside the screen, dump the armload. Lope to the kitchen for special incendiaries such as waxed food packaging or an empty milk carton. Shove on a backlog, put down paper and carton, arrange the kindling (a little more than is really needed), scratch a kitchen match, light the paper in as many places as the match allows.

At this moment the tempo is no longer *prestissimo* because even an infant fire has a say in how fast it is willing to spread and grow. If the exploring little flames seem to pick through the kindling in too deliberate a way, it is all right to ball up a sheet or two of newspaper and ram it beneath. (Lighting the early-morning fire is the one occasion you never stop to glance at a headline or advertisement in the paper before crumpling it.) Add more wood, heavier kindling, a husky branch or two. But be wary of adding too much too soon, for a cub fire lacks the thermal momentum of an adult blaze and can be retarded or choked from overfeeding. Pull up a stool to the hearth and spend a few minutes as the fire develops by tonging ends of burned-through kindling into or directly under the core of the blaze. Now new warmth begins to radiate to your hands and bare shins, too weak to make inroads on the chill of the whole room but already enough to still the chattering of your teeth. By the time it is right to add a third piece of full-sized wood, bridging the first two but cocked just enough to let the flames speed upward, it is also right to make a foray to the kitchen to put the coffee on and to rummage for a prebreakfast doughnut or roll.

By now the cool-morning fire is virtually as delightful as the one imagined in bed a few minutes before. Its radius of comfort is small but growing, and a rocker can be pulled close so that slippered feet can be propped up, deliciously warming the underside of the shanks, and the ash from the first cigarette can be flicked into the fire with scant effort. The steep temperature gradient, almost too hot in front and still chilly in back, is something not often encountered in the world of central heating, and it is easy to see how it annoyed Ben Franklin, who, reasonably, objected to it as a permanent condition of winter life indoors. Our reference is different, since we are accustomed to central heating having a diffused evenness quite unknown in the 18th Century, or for that matter almost any time before this century. For us the morning fire in a cold room seems a little strange, the temperature differential having the effect of enhancing the fire-ness of a blaze, brightening its colors and dramatizing its warmth, rather in the way color film sometimes exaggerates perceived reality. It is in a cool room that infrared radiance from a fire is most evident; your senses tell you almost instantly when someone shadows you by stepping in front of the fire.

As a morning fire rounds into maturity and the rocking chair has been inched farther and farther back, it is easy to forget how chilly the rest of the house remains, although a trip to the kitchen for a second cup of coffee will be a reminder. This leads to one of the greatest amenities provided by a morning fire: you can dress in front of it. There is no need for the Spartan discomfort of putting on an icy shirt, damply cold from the drawer, when it can be brought directly to the fire, held up in the radiance until the chill disappears from every fold, and then slipped on in luxurious comfort. In a cold house, putting on a prewarmed shirt provides a pulse of pleasure that is as sensuously enjoyable as, say, a great stretchy yawn.

Other daytime fires are generally less flamboyant, less directed toward immediate heat. Often they are woman-built blazes kindled in late morning or afternoon to add a comforting splash of color and warmth to an overcast wintry day. Such fires are also useful to welcome guests; a fire is inherently hospitable. While women can be deft about the hearth, it is observable that their fires are frequently different from men's, tending to be smaller, more centralized, and somewhat more precariously constructed. (When etiquette permits, most female-built fires want to be kicked a little, and then built up.)

Fires built in the evening also lack the pre-eminently physical appeal of early-morning fires, tending to be more subtle in their uses and effects, less solely directed to the sensorium. Consider for example the fire by which late-evening reading is often done. It is usually a mild-mannered middle-aged fire, companionable but unobtrusive. Without thinking much about it, its firetender is likely to string it out by adding one modest piece of wood at a time, putting it on a bit late so the fire can putter along only in a slow agreeable way. If the book is absorbing, able to inhibit those slivers of consciousness that can break free and wander about on their own during reading, the fire may burn so low that it is necessary to stop and build it up again. Such a fire is plainly not prized for warmth nor visual show nor the induction of hypnotic reveries. Its rôle is to supply companionship when everyone else in the house has gone to sleep, to be a comforting presence, rustling a little, like a friendly dog snoozing near by. Its companionship is not simple, however, being in some fashion related to a submerged awareness of night and darkness outside. If the night turns stormy and rain drums on the roof until the downspouts gurgle, or if sleet dashes against the windows, the night-reading fire is immediately built to a bright blaze, an instinctive defiance to the storm and darkness outside.

Evening fires not only provide quiet companionship but also are superb adjuncts to it. A fireside is an excellent place to renew an old interrupted friendship or, as a practical if delicate matter, to see if you want to renew it. The best friend is the one whom you sit by a fire with most easily, talking the hours away or, if the collective mood is otherwise, enjoying long easy silences. Thoreau, not the most sociable of men, observed that fire is the most tolerable third party. Nowadays it is usually held invidious to examine the elements that compose a friendship, a contemporary taboo possibly based on the Heisenberg principle that to measure something is to risk changing it. But if circumstance ever requires you to consider a friendship analytically, a recollection of how you sit by a fire with someone is insightful.

One fascinating fact about the companionship that exists before a hearth is that it remains singularly clear in memory. When I think of the people I have known well that are now dead—a goodly company of vivid human beings, remembered in a hundred settings, with highly individual voices and gestures and postures, all of them talking and laughing, but frozen in time now as in the children's game when the music stops, never to grow a day older although I do so every day—I am struck by the fact that those most sharply recalled are recollected by firelight. This is curious, because firelight is not the best illumination for the capturing of sharp images. Of course fire is a natural metaphor for life, one that has been long used by peoples all over the world, a usage deeply embedded in language: *spark of life, dying embers, dead ashes.* Probably some dealer in psychiatric magic and spells can argue that the phenomenon of firelight recollection is merely an illusion arising from an elegaic mood interacting with the natural metaphor. Not liking so glib a dismissal of my friends, I have constructed a countertheory, which is that some characteristic of firelight can create holographic images that are stored in some special cerebral vault, a particularly protected file of memory. I have taken pains never to discuss this theory with any physicists, however, feeling certain that they would reject the idea in their recklessly cocksure way.

There are numerous other evening fires, each subtly different in construction or management: fires to have dinner by, to play chess or cards by, and to listen to music by. There are fires of celebration, reflection, and solace; fires proper for Christmas Eve, New Year's Eve, and Twelfth Night; fires for snowy evenings when drifts sift deep outside and others for windy starlit nights of great cold. There are fires to talk and laugh by and others to wait up alone by; fires to make love by and fires to sleep by. It is notable that, fires being among the most ancient human civilities, few seem adapted to ill-tempered quarreling by. These differences in individual evening fires are those that a skilled firetender, drawing on the resources of a well-stocked woodshed, can instinctively create, differences in scale, tempo, and stability. His concern is to match the fire to the circumstance and need. It is inconvenient, for example, to make love by a fire that morosely dies, or else roars, spits, and then incontinently rolls out smoking on the hearth.

¶ THE STATUS of the fireplace fire in the closing decades of the 20th Century is difficult to assess, for it is now so widely esteemed as a domestic ornament that it has become coated with a glaze of candied sentiment. Greeting-card racks abound in representations of homey fires, and at Christmas the pictures of pretty flames multiply beyond counting, even if you disregard those in which the fireplace is a prop for Santa. Firelight flickers in many a gluey love lyric, and is a favorite with illustrators, calendar artists, and others employed in idealizing daily life. Real-estate salesmen and architects are aware that a fireplace is needed for something mysteriously described as "gracious living." There is even the sanction supplied by counterculture attack, as when in 1973 a little band of ecofreaks picketed a New York City apartment house advertising wood-burning fireplaces among its features.

For years interior decorators, employing the peculiar prose with which they ply their trade, have pronounced the fireplace a natural focus of interest in the sophisticated management of space, color, texture, and function. Fireplaces, we are instructed, are vital adjuncts to conversation pits. If you examine photographs made to capture the decorator's work—a fascinatingly unreal art form in which all evidences of human presence, such as sofa cushions that have been sat upon, newspapers and magazines that have been allowed to collect in unaligned piles, and ashtrays that have actually received ashes, are rigorously suppressed and replaced by an eerie orderliness, as if the quarters were occupied only by bodiless wraiths—you will frequently observe that the decorator's photographer has specially created a bogus fire for the occasion. This is typically a newspaper crumpled beneath the stagey birch logs that, lit and captured in blurry fashion during the two-second exposure, generate a heatless, ashless, emberless blaze to warm the wraiths.*

Commercial exploitation of the fireplace as symbol does not take place in the abstract; the taintmakers do

* Collectors of fake fireplace fires can find them everywhere, typically in magazine advertisements for furniture and carpets, and now increasingly in television commercials. Here the fire has become a conventionalized symbol for home at its most intensely homey, powerful for the marketing of life insurance, dog food, and the special felicity said to result from keeping a laxative at the ready. The fireplace has become a visual cliché, almost as divergent from reality as those strange objects—conch shells, glass jars, and statuary (it used to be all Nefertiti or Ftatateeta but pre-Columbian figures are coming on strong)—that advertising illustrators perplexingly place on bookshelves in place of books. But it is well not to snort at such symbols, for a wise Providence will take care of them. Thus the proud smokestacks that once ornamented industrial letterheads would now simply be obtuse; thus the bust of Napoleon, once a signal that its owner was a deep thinker, is now a sign that the poor man has gone loony.

not create attitudes so much as live parasitically on pre-existing ones. Clearly many people are profoundly responsive to fireplace fire. It gives not just warmth but also a magic protection against the encircling dark. It is a joy to watch, a ballet of light and color dancing above hypnotic embers that conduce to gentle trance states. A fire, unlike many pleasures, is largely guilt free, and it is egalitarian, shedding its beneficence on all without regard to station or wealth. Fire offers rewards to all ages and both sexes: children are fascinated and sedated by it; women, being set designers at heart, are sensitive to its enhancement of room, home, and complexion, and men dream by it and also relish its need for technique, rewarding deft manipulation by poker and tongs. An open fire is part of most current conceptions of the good life.

This last attitude is in considerable degree a very recent one. For 99 percent of the 750,000 years since man domesticated fire, it has been thought of as a necessity rather than an amenity. Central heating, like bathtubs and inside toilets, has been everyday technology for only about a century in much of the world's Temperate Zone, and still hasn't spread in immense areas that are exposed to chill winter seasons. *Automatic* central heating is even newer; some kind of daily firetending was essential in most households until very recently indeed. My father, coming of age approximately with the 20th Century, tended a coal furnace twice a day during most of his life, invariably careful to keep coal or ash smudges from his starched cuffs and detachable collar. As he rocked the cast-iron grates (a sound telegraphed throughout the house by the steam pipes), broke up clinkers, shoveled coal with accuracy into the firebox, scanned the water gauge, and set the draft and damper controls with the same nicety that his children now use to tune color TV, he gave no evidence whatever of feeling inconvenienced by this twice-a-day chore of serving the big iron boiler in the basement. I suspect that if he thought about it at all, it was to admire its modernity and efficiency in contrast to the home of his childhood, where a fireplace, a kitchen range, and an enameled parlor stove generated uneven pockets of warmth against the bitter northern winter. He was a disciplined and equable man, as well as a deft firetender, evincing only a small glint of annoyance when he came home to discover that my mother, aunt, or the cook had tampered with the furnace controls in the afternoon. Women, he once confided in me, tend to confuse dusk and chill.

¶ WHEN Miss Hewins told her second-grade geography class about the Temperate Zone, we felt that it was an appealing idea to have felicitous bands about the Earth where people led happy and productive lives midway between steaming jungles and ice floes dotted with polar bears. Miss Hewins is inaccessible now, but I have often had the impulse to point out to her that, thanks to the tilt of the planet's polar axis and to a varied assortment of global and continental weather phenomena, the Temperate Zone is a region where for part of the year for part of some days, the climate is temperate, being in general abominable the rest of the time. Although mid-continental American Indians had fire at their disposal, their technology was not extensive, and it seems reasonable that, Temperate Zone or not, they spent so much time being miserably uncomfortable as to encourage the growth of the stoicism for which they were noted.

Certainly getting through the winter without freezing to death or starving was a desperate concern to the first European settlers in New England. Those who managed it emerged from a sieving process which, for a time at least, guaranteed that the survivors were biologically select. In a settler's hut, fire was far from an amenity; it was as essential to survival as food and water. Nearly half the Plymouth settlement died in their terrible first North American winter from cold and starvation, and the Indians were moved to pity and aid for the strangers who had blundered upon their shores.

It is a wry circumstance that some descendants of these settlers have constructed an ancestor worship that seems less concerned with hardiness than gentility. They see themselves sprung from persons who quaintly christened their daughters with names like Patience, Humility, and Prudence; who stepped ashore to construct saltbox houses of outstanding charm, and who shot Thanksgiving turkeys while dressed in pageant costumes with silver shoe buckles. While these innocent conceits are harmless, it also seems reasonable that—to the degree that a fifteen- to twenty-generation span provides grounds for genetic vanity—the company of revered ancestors could be broadened to include all who arrived in America in the first half of the 17th Century, and to respect them more for bravery than gentility. The courage, resolution, and energy needed to make a life in a howling wilderness is quite enough.

Fire made from flint and steel, along with the axe, adze, fishhook, and blunderbuss, supplemented by practical competence at food-gathering and food-planting learned from the compassionate Indians, made possible the first precarious European beachheads on the North American continent. Not all settlements survived; the record from 1607 to mid-century has many accounts of pitiful groups that didn't make it through the first winters, and it was the lucky settlements where survivors could take ship for a less hostile spot. It is fascinating to compare the unpreparedness with which these little parties of innocents arrived in the New World with the sturdy competence they were to show only a few years later. They came armored in ignorance, displaced persons in the religious conflicts of their century, believers in the fanciful accounts of ease and plenty told by holders of royal patents and incorporators of the plantation companies, themselves perhaps not accurately informed. As a consequence the first expeditions were ineptly planned and stumblingly executed. After a series of delays and mischances, its food stocks running low and its navigation faulty, the *Mayflower* finally put its company ashore at Plymouth on the first day of winter, hardly a rational time to begin living off the land in New England. The first dwellings ashore, imitations of Indian long houses, were flimsy sheds of saplings and poles, covered by bark and reeds, entirely penetrable by winter winds. Their first fireplaces were circular hearths beneath smokeholes in the roof. In Connecticut and in New Amsterdam first settlers arrived in better season, before the ground was frozen rock hard, and they simply dug holes, rectangular cellars lined with logs or mud-mortared rocks, and having thatched roofs just above ground level. It was in these wretched hovels, around roaring fires—wood was one necessity in abundance—that, half starved, often sick, but with singular spirit and energy, these people contrived the first permanent lodgements on our shores. Did they sing as well as pray around the fires in their smoke-filled huts? How did woodcutters and hunters protect their hands and heads against the astonishing cold? Their death rate was appalling, and particularly high for the women.

Matters grew better, or at least less catastrophic, at a rate that reflects the resourcefulness and adaptability of these people. It is in dwellings and their fireplaces that, although details are tantalizingly scant, this improvement can be traced. The flimsy huts were quickly replaced with solid buildings, constructed in New England of heavy timbers adzed to fit together and chinked tight with clay. Only seven years after the landing at Plymouth the thatched roofs familiar from the old country were prohibited: much too dangerous above the great roaring fires that the new land demanded. A reflective fieldstone fireback appeared by the hearth, and the primitive smokehole gave way to a wattle-and-daub smoke hood, a kind of rudimentary chimney.

For a time a technological obstacle prevented true chimneys. Stone was available in abundance (except in Tidewater Virginia, which as a consequence resorted to brick), but clay with straw or hair binders was an inferior mortar, especially with the rounded rock that the glaciers had left behind. Clay or even mud was all right for wattle-and-daub buildings, though it needed constant replenishment and repair after drenching weather; and for a time clay-lined wooden chimneys were attempted. These too had been known in England, satisfactory above modest cooking fires, but like thatched roofs they proved perilous above great fires blazing day and night. Soon an unidentified genius solved the problem. Limestone might not yet have been found in the New World but seashells there were aplenty. Ground to powder and roasted, seashells made a cement from which an excellent heat- and water-resistant mortar could be compounded. Soon durable fireplaces of almost imperial proportions were being built. They were thermodynamically crude—scarcely a serious drawback where cleared land had to be wrested from primeval forests—but they also wonderfully combined heating, cooking, and lighting facilities in a core structure that anticipated by centuries the utility-core concepts of modern architects. Under the stress of necessity-powered inventiveness, the fireplaces came to have devices of growing sophistication for roasting and boiling, frying and grilling. Rising broad and tall above stone hearths, those fireplaces provided the warmth, nourishment, and firelight that must have much comforted those driven souls.

Another and related example of necessity-sparked discovery occurred very early. The first European settlers in America were true representatives of Iron Age man, and they set about finding and refining iron ore with astonishing promptness. In Virginia a primitive ironworks was established as early as 1619, its production being periodically interrupted by Indian attacks. It was in Massachusetts in 1643 that John Winthrop Jr. and others founded The Company of Undertakers for the Iron Works, producing bog iron from the limonite found along the Saugus River in Lynn. The deposits were modest and the refining process crude, problems that were trivial in the light of how much native iron was in demand, not solely for tools, nails, and simple hardware but even more for the kettles, pans, and cranes that each new fireplace required. Only a generation or so after predecessors had huddled in their first hovels, a rapidly increasing population was established in new little towns along the Atlantic coast, each dwelling enclosing a bright and comfortable cooking fireplace.

¶ IN THE DAYS when fireplaces were not just amenities but absolute essentials to winter survival in the Temperate Zone, it was natural that they were the subject of concentrated attention and thought. Some aspects of this thought, and some of the brightly imaginative minds that focused on fireplaces, will be examined elsewhere in this volume, and it is sufficient here to suggest the range of contexts of this thought. There was, early, the fixed gaze of the technically inventive, beginning with ingenious solutions to the problems of hearth cookery, and extending as invention often does to the boundaries of the gloriously queer. Trammels, cranes, kettle-tilters (as clever as any gadget you can find in Bloomingdale's), griddles, trivets, and fire-carriers (no need to spend long patient minutes with flint-and-steel at a second fireplace when you can carry to it a nucleus of existing fire) were among the scores of ingenuities created for the fireplace. A few conundrums have also survived: devices not now understood, for the solution of unknown problems. Uncounted men have grappled with the problem of slowly rotating the skewered meat to be roasted in the rich infrared radiance of a hardwood fire. It is not arduous labor but tedious, with burnt meat to penalize the easy sin of inattention when the task is assigned to a slave or a child. One solution re-invented repeatedly was the smokejack, an impulse wheel of iron, placed in the column of ascending heated air within the chimney, and rigged by gears or chain to turn the spit. This was a sound enough idea, though not simple to execute with blacksmith technology. It competed for favor with the steamjack (a teakettle blowing on a primitive turbine wheel) and the clockjack (a pendulum-timed escapement, driven by a suspended weight or rock). There was also the dogjack, powered by a small dog, walking a treadmill and rendered especially amenable by the delicious smell of cooking meat.

Not all the inventors were gadget-minded. Two of the most original scientific minds of the late 18th Century, Count Rumford and Benjamin Franklin, were fireplace theoreticians. The Count, a wonderfully versatile and faintly raffish fellow, was among many other things a pioneer thermodynamicist who helped bring the physics of heat out of its ancient welter with a mystic substance called "caloric." At one point in his career the Count was a doctor of smoky fireplaces with a fashionable clientele among the English upper classes. His idiosyncratic fireplace designs have never been excelled in radiant effectiveness and disinclination to smoke. Franklin was his contemporary and, in respect to fireplaces, his rival, the author of shrewd writings on the prevention of smoking chimneys, and inventor of the celebrated Franklin stove/fireplace, presented royalty free to all mankind. The Franklin stove moved the fireplace out into the room and out of its prior dependence on radiant heat transfer alone. Franklin was the herald of a kind of Iron Age on the hearth: the United States Patent Office in the period from 1790 to 1871 issued patents for more than one hundred different varieties of iron fireplaces.

Interwoven with all these efforts were wholly different products of human concentration on the fireplace. An important one was the awareness developing all through the 18th Century that, a fire being inherently pleasing and attractive, its place ought to be made handsome too. Brass, one of man's most friendly and manageable alloys, was perceived to be wonderfully sympathetic with firelight, ideal for elegant ornamentation of fender, screen, and the vertical portion of andirons. Hearths grew to have proscenium arches framing the theater of fire, with paneling, columns, and mantels of increasing elaboration and delicacy. Fireplaces displayed more than the separate skills of the master mason and brassfounder, and the deft talents of cabinetmaker, joiner, and woodcarver; they also became a major concern of the leading architects of the century, a fit subject for thought about location and size, exact proportions and the nicety of fine detailing. Toward the end of colonial times in America and on into the Federal period, fireplaces were built that were and are aesthetic achievements of high distinction. It is no denigration of their beauty to recognize that they were also valued as a means of demonstrating an owner's rank, wealth, and esteem.

A second variant theme entwined in fireplace thought was, of all things, health. It seems an odd correlation today, at least until we recall that scientifically coherent understanding of the etiology of disease is only about a hundred years old, and that long before germs and viruses were identified there were mephitic and miasmic vapors of malign power. It was clear that fireplaces were intimately involved with air and air was risky stuff, often impure if not noxious, the source of everything from the rheumatics to quinsy, palsy, and consumption, a notorious source of catarrh, rheum, phlegm, as well as being the essence of the drafts that could so easily carry off the infirm. As a consequence of this general view, the fireplace was believed to be directly influential on the well-being of those who lived by it. No close analogue of these anxious beliefs exists today, except perhaps in some degree in the magic imputed to the foods we eat.

Finally, there were economic and occupational aspects to the fireplace, which had an importance that is now easy to overlook. The fireplace not only made life possible for everyone but also supplied a livelihood for many. Woodcutters and sellers of firewood were only the beginning of it: coal miners and coal dealers (as cheap firewood became scarcer), masons, bricklayers, carpenters, andiron-founders, brassmongers, purveyors of marble, tile, and firebrick, importers of ornamentation, carvers and gilders, smiths specializing in firetools, repointers of mortar, and chimneysweeps and chimney-viewers. The viewer was an opposite number to the fence-viewer, whose responsibility was to see that hogs and cattle did not escape to root up crops, and both were officials appointed by the community in the 17th Century.* A chimney-viewer was a formidable fellow with authority to condemn chimneys so defective or choked with soot as to be likely to take fire and spout sparks on windy winter nights, endangering nearby roofs and risking general conflagration. His appearance was alarming, particularly if he wore a long face, for his was the absolute authority to forbid the use of a fire. Partly

* The presence of these officials underlines the extent to which early settlements were communes, a view not cherished by those who prefer to think of them as pockets of admirably untrammeled free enterprise. A small settlement newly established on the edge of a wilderness obviously had shared collective interests in many areas, including protection against famine or disastrous fire. Fence-viewers still exist today in some corners of New England that find comfort in older ways; chimney-viewers have been transmuted into fire marshals and insurance inspectors.

as protection against this powerful civil servant, chimneysweeps came into widespread use in towns and cities. Numerous efforts to clean out chimneys by nonhuman means were attempted—thrashing long chains inside, dragging fresh-cut brush through, even dragging through indignantly honking and wing-thrashing geese—but none had the ability to clean, report, and repair that was possessed by an agile small boy. Apprenticeship often began at age eight, with lads who were small for their age preferred, and was terminated when the lad grew too large to fit, if indeed he was so fortunate.

Although today this band of fireplace-related occupations has changed character, and although the fireplace itself is growing increasingly to be an adjunct to the good life, fire is still the means by which virtually everyone dwelling in the Temperate Zone manages to get through the winter. This remains true whether the fire is in the fireplaces and stoves—metal-boxed fireplaces—that are still the only source of warmth in many rural houses, or whether it is in the modern hypocausts, the scary furnaces roaring unseen within the great dwelling hives in our cities. Technically, there exist two vanishingly small exceptions to this generality of fire dependence. One consists of those who might winter through on the basis of electrical heat generated by power from nuclear reactors. This exception is rendered theoretical because contemporary electrical grids mingle the power from nuclear and fossil-fuel plants, and there is no convenient way to examine a volt-ampere to see if it came from fire or fission. The second exception is legitimate but insignificant, consisting of the small number of people who are warmed in winter by heat pumps. A technological oddity, the heat pump works by collecting and concentrating heat from some nearby natural resource that in fact isn't particularly warm, such as well water or deep lake water. The heat pump is elegant in concept but (as I discovered during the one winter that a heat pump kept my family warm) disquietingly inelegant in present practice. While it has no fire within, it does have so many pumps, compressors, and blowers that it hums, gurgles, and whooshes through the long winter nights as it performs its witty thermal card trick.

For all of us who live beyond the equatorial savannas where man was first differentiated, fire remains an essential to life during at least part of our planet's annual orbit. This has been true ever since the remote time when man became the most extraordinary of all the animals on Earth, when fire became the semimanageable magic that made him Man. If awareness of this is buried deeply in the collective memory of our species, then it is less strange that we stare at a fire's luminous embers with a sense of deep belonging.

THE SPARK

When a tough question needs answering, it can help to call in a mythmaker. It is an approach that has been very useful in accounting for the domestication of fire. Prometheus, the Titan who stole fire from the gods and gave it to man, incurring Zeus's formidable wrath, has been the beneficiary of the special prestige conferred on figures in Greek mythology; but he was very far from being the only legendary fire-bringer. All over the world primitive people have sat by their fires, staring at the leaping flames and glowing coals, wondering how fire came to their ancestors. Folk tales have been spun in every land about the hero, or spirit, or supernatural animal who performed the feat of borrowing fire from the Sun, or stars, or lightning, bestowing it as a priceless gift on those who came after. By the number of legends alone, it is clear that the domestication of fire has been puzzled about for years beyond counting. There is an almost universal feeling that something brave and remarkable was done.

Let it be said at the outset that no one now knows how men first learned to make use of fire in the misty ancient times. All we have are guesses, nonpoetic myths. The preferred source of explanation in our day is not so much the tale-spinner as it is the scientist, in this case the anthropologist. His is a relatively new discipline, little more than a century old, and still making long strides in data, technique, and theory. And while it is in the nature of all oracles, science included, to receive direct questions as bad manners, worthy at best of an oblique answer, anthropologists as a class are not insistently unintelligible, and are willing to admit that the domestication of fire by man—no other animal has come close to managing it—is in fact a perplexing problem.

For one thing, there are no plausible transitional processes, none of the ramps of learning by which we can imagine that other achievements were managed. The hand axe, spear, and flint tool offer nothing like the problem. Fire is a discontinuity, a chemophysical reaction that, for ancient man, had no transitional phase. He could not have practiced with almost-fire. Moreover, his mastery of it is the more remarkable because, at the beginning, fire was very frightening to him. Almost all animals seem to be endowed with a profound fear of fire.

Not only do we know nothing about how the extraordinary feat of bringing fire within the cave was first managed, but we almost certainly never will know. Considering the ephemeral as well as ambiguous nature of any likely evidence, the record is surely lost forever. We can fix this extraordinary event only in one dimension, time, and the date proves to be astonishingly early. Fire was first tamed at least 700,000 to 750,000 years ago. This means that this feat of courage, discipline, and high intelligence was not performed by *Homo sapiens*, the animal that is us, but by his distinctly unprepossessing predecessor, *Homo erectus*.

Several ancient sites give testimony. In 1960 highway workmen blasting a rock cut at Escale, in southern France not far from Marseilles, broke into a deep limestone cave. Within, on a cave floor dated as being approximately three-quarters of a million years old, anthropologists found evidences of ash and charcoal, fire-cracked stones, and five hearth circles each about a meter in diameter. This is the earliest hearthsite yet known. Along with a similar but somewhat older campsite at Vallonet—which included stone tools, broken animal bones, but no hearths—these are the earliest signs of man outside his birth-continent of Africa. The evolutionary stage of the creatures who left these poignant traces has not been identified. They just might have been exceptional members of the Australopithecus family of hominids, a precursor species that had been developing in Africa for many millions of years and already a toolmaker. More likely they were *Homo erectus*, a bigger-brained species first appearing about 1.3 million years ago. Perhaps stimulated by weather change and by restlessness, Erectus had begun to wander beyond Africa into Europe and Asia.

The next early fires burned much later than at Escale, although still almost a half million years ago. They were kindled many thousands of miles away at a place called Dragon's Hill, which is at Choukoutien, about thirty miles from Peking. The site is a rich one, much investigated by anthropologists, and a complicated record of cave occupancy over a span of time has been partially deciphered. Ashes, hearths, and burned animal bones suggest that the creatures dwelling there nearly 500,000 years ago were accustomed users of fire. Parts of fourteen skulls have been recovered from the cave, with enough other human bones to suggest that at least forty individuals died there; it was a time before man buried his dead. Peking Man (currently identified as Erectus) was a tool-using carnivore, a hunter with a taste for venison. Nothing is known with certainty about the steps that preceded the appearance of fire in his lair. At one period there were simply established hearths in use, rather as though to have fire in the cave, so convenient for light, warmth, and the cooking of deer, was natural for Peking Man. (Fire in fact served him in other ways, as we shall see in the next chapter.) There is simply no evidence of how he or his predecessors managed to domesticate fire, one of the most fateful steps that man has ever taken. Although evidence is lacking, we can fit reasoning to probability, and devise hypotheses of what might have happened in the distant days when no campfires gleamed in the dark anywhere on the planet.

¶ IN SPECULATING how this technical achievement may have come about—a pleasant armchair sport, nicely suited to pondering before an open fire—it is well to remember that *may* is not remotely interchangeable with *did*. You may imagine what you want, but actual events have a way of occurring in close agreement with the laws of probability. Thus you are entirely free to hypothesize that fire was first discovered by an exceptionally gifted hominid named Pyro—it could have been a female named Candens, if your quirks run that way—who one fortunate day was absently rotating a stick in a manner later used with firedrills. Then Pyro, or Candens, delightedly observed that a wisp of smoke, and a charred area, and then a tiny glowing kernel magically appeared at the point of friction. If you wish, you can construct a different hypothesis in which Pyro, or Candens, was absently whapping a piece of flint with a hunk of iron pyrites, and became enamored with the evanescent flicks of light that sometimes resulted. Then he or she discovered with high intelligence and luck that the flicks could be caught on a piece of tinder, blown upon with delicacy, and nursed into a self-sustaining chemical reaction that was extremely handy to have about the cave.

Should you cherish either of these hypotheses, you may enjoy the thought that no one will ever be able to convince you to your satisfaction that you are preposterously wrong. The truth, of course, is that these scenarios are unlikely in high degree. Firedrill friction and flint/iron sparks have been used by men to kindle uncounted fires, but neither is suited to the concept of accidental discovery. Each is a firemaking system made up of moderately complicated and extremely technique-sensitive procedures, involving a series of critical steps. Each method is demanding enough to have made occasional problems for people who knew precisely what to do, using precisely the right materials. So it isn't plausible to imagine Pyro or Candens achieving their miracle through a series of felicitous accidents. A more fundamental defect is the assumption that man first learned to start a fire and then discovered how to manage it. A reverse sequence—in which the skills needed to sustain and control fire ignited by natural causes came first, and were learned incrementally over a long period of time—has much to recommend it. The tricks of ignition might well have been mastered many millennia later. It is intuitively plausible that *Homo erectus* would have found this a much easier path to the summit of his incredible achievement.

As to where pre-existent fire could have come from, a lively imagination provides answers by the dozen. Lightning and volcanic eruption, obviously. A rockslide can create sparks. A heavy tree branch partly broken free and lodged in a crotch might work in the wind until the friction point smolders. A naturally caused brush or forest fire might ignite a semipermanent fire of the sort occasionally known today, a fire that fed upon peat, a coal seam, or oil seepage to burn smokily for years, unquenched by rains. On a small and geometrically neat scale, a clear icicle of the correct convexity might have focused sunlight on dry wood. Slacken the reins on your imagination and it becomes a game. Here's a flaming meteor, incandescent from its hypervelocity passage through the atmosphere, that drops plop at the feet of an exceptionally intelligent hominid, a Newton or Einstein in animal skins. Over there in the clearing are the favorite *dei ex machina* of our day, the extraterrestrials who have emerged from their saucer to present matches and Zippo lighters to the excited hominids clustering around, as starter sets on the road to civilization.

A difficulty with hypotheses constructed on bizarre events is that they are intuitively unsatisfactory. They are inelegant and faintly bogus, like the questions asked of a lecturer designed less to win information than attention. They violate the rule that the simplest is the most plausible. You may hypothesize a band of bright, fire-ready hominids that chanced to come to a neighborhood of volcanic activity, perhaps close by one of those smoking conical peaks encountered so often on late-evening television. Immediately the hypothesis is in trouble. A volcano is unfriendly: it shakes the ground, smells bad, and makes hostile noises. So the first problem is to devise explanations of why the bright hominids didn't immediately depart the neighborhood, as any sensible life-loving animal would. A second problem is that you are constructing a special-case hypothesis—continued lingering around alarming seismic events—and the special condition is simply not as satisfying as the general one. It is true always that almost anything could have happened, but what did happen was dictated most rigorously by probability. It is not easy for us to apply the distribution curves of probability to behavior, caught up as we are with some identification with our species, and aware that its actions can be idiosyncratic to the edge of nuttiness. When you discuss the matter with a technical man, a physicist, say, who is accustomed to considering the behavior of unimaginably large numbers of unimaginably small particles, some of them very queer little entities indeed, you come away with heightened recognition of the power of probability to predict events in the known universe.

Certainly volcanism could have been the ignition source for man. A smoldering underground fire is another real possibility, for primitive man might have dwelt near by and become accustomed to it. But statistically lightning is by far the likeliest source of natural fire for man. About 44,000 thunderstorms occur about the world each day, and there is an average of 400 lightning bolts occurring somewhere each second. One early unmanned satellite, OSO 1, radioed back the information that lightning strokes to the surface are not randomly distributed, but occur preferentially over land. (This was not at all what little OSO 1 had been sent into orbit to look at, but it had already done its regular job and plainly couldn't resist a serendipitous tidbit.) When lightning strikes a tree it can be formidably energetic, pouring tens of thousands of amperes through a small path, exploding wood fibers by the generation of superheated steam, and scattering pieces of burning bark and wood over a wide area. Foresters know that, despite Smokey the Bear's misapprehension that only man is vile, lightning starts more than 7,000 brush and forest fires every year.

So brush and forest fires were not unknown to ancient man, nor were the racing grass fires that can flare up on prairie and savanna. Erectus was surely terrified by them, by the strange smell and smoke, and then the crackling, roaring flames, as implacable as tiger or wolf. Such fires must have herded men, driving them in something of the panic that possesses other animals. Something like, but not quite: we can speculate that in man fear did not totally dominate his mind, but left a little room for observation, analysis, a choice of strategy. Moving downwind was basic but an oblique course could be better. Tree-climbing was worthless but going to cover in a cave might help. The terrifying flames did not pass over bare rock faces or big sandy areas, and swimming out into a fair-sized pond was a way to safety. Fire left behind a blackened and smelly countryside that was worth scouting if you could get around the red flames because fire did not come back to pursue you there. Slowly, after uncounted generations of men, it was learned that behind a fire there would be fire-killed game that could be eaten, even though charred. (It was necessary to be careful not to step on or touch small glowing pieces of fire remaining after the roaring wave of flames passed on.) In time it became known that fire wasn't the same absolute death as being caught unaware by a tiger, but even had a good side in that it could provide a full belly. From the discovery that fire was a mitigated evil it was only a step to the discovery that fire could be an aid in hunting—that game could be killed more easily during panicky flight.

No way is open by which we can enter the minds of predecessor men and comprehend the forms of their thought. We don't know, for example, if fire was first personified as a devil god, or if it was a retribution imposed by angry spirits for impious behavior, or if it was simply one more damned chancy event in a precarious life. The thoughts that coursed within those big-brained, thick-walled skulls are forever lost to us. The first opportunity to glimpse abstract concepts in predecessor man occurs with men who lived hundreds of millennia after the initial domestication of fire, when ritual burials began to find use. To bury one's dead with food, tools, weapons, and possible badges of rank is to imply a belief in the possibility of life after death, and of other worlds than the one departed from. The burial of an occasional object with the dead man might be interpretable as an act of affection or reverence, but the extensiveness and frequency of ritual burial clearly suggest the concept of resurrection to an afterlife. Such burials first appear about 60,000 years ago. Before that the dead were simply discarded, or disposed of; afterward, there were mute signals of the presence of ideas that have beset or comforted men ever since.

Our probability-based speculations about how fire may have been domesticated have brought us to a time when fire was known and in slight degree used. It seems reasonable to imagine that the next step was taken by some gifted and courageous individuals, very possibly youngsters. Curiosity is a trait displayed by many animals, particularly in man, and among men most strongly in the young. Curiosity in animals other than man seems to be generalized and rather detached, an interest in something unusual in appearance or smell or sound. But in human beings and particularly young ones—whose brains acquire new information about the world at an exceptional rate—curiosity leads to a kind of participatory investigation, beginning with looking, listening, and smelling, but soon moving on to touching, probing, and experimenting. Children are explorers of their environment, and perceive reality with a freshness and alertness that adults often lose. Our hypothesized fire-investigator was probably not a small child, however, because of the high courage his actions called for; his daring and curiosity had to be sufficient to outlast a short attention span and outweigh innate fire fear. Perhaps he was fourteen or fifteen years old, acutely observant, a natural experimenter.

We don't know how much speech he had. Probably not much, but nevertheless he must have been traveling the long trail that led from the simple pack cries of hunting toward a supple communications system able to deal with ideas. He was a toolmaker, with handaxe, spear, and edged stone tools for cutting and scraping. Watch as he cautiously approaches a windfall tree trunk smoldering on the ground after a brushfire has swept past: circling, tense, jinking a little, careful not to step on anything still smoking, tempted to cut and run, but still wholly fascinated by the fire flickering at the side of the trunk. It might have seemed to him a cub fire, a puppy blaze, relatively innocuous in size and youthfulness. Watch him poke at it with his pointed hunting stick, and then look wonderingly at the even smaller fire that has magically appeared at its end.

This was truly dangerous business, with high risk of severe or fatal burns. Visualize how one of us might react if presented with fire never before experienced, or if—more appropriate to our time—confronted with several substantial pieces of plutonium. We would, most of us, come to the situation with unstable oscillations between timidity and rashness, with a real peril of injury. It seems possible that many of our ancestors died in learning to manage fire, most of them painfully indeed, with the exception of those who lost their lives to carbon monoxide and asphyxiation in the process of bringing fire into shelter. The taming of fire must have seemed baffling—as does research into any new field until underlying principles are mapped and generalized—full of traps for the unwary even as fire became partially manageable. Unknown ancient men were predecessors of Louis Slotin, a man who died a strange death in 1946 at Los Alamos, one of the first peacetime victims of nuclear fission. In circumstances that seem uncomfortably casual a few decades later, and that were even then known as "twisting the tail of the dragon," Dr. Slotin was at a table demonstrating weapons assembly when for an instant he accidentally brought two subcritical pieces of plutonium too close together. The instantaneous blue Cerenkov glow, a new kind of flame, signaled the presence of intense radiation amounting, it was later calculated, to more than twice the lethal dosage. As quickly as human reactions could do so, Dr. Slotin separated the pieces; he went off to the hospital and died nine days later. Something like this, in elements of eerie and mysterious fatality, must have taken place many times in the ancient conquest of fire.

But failure is not as much evidenced as success, and some men succeeded. A large body of learning grew from that first spear poked so hesitantly at the burning windfall. Doubtless the knowledge came in stages, the product of many investigators. Fire transfers itself from burning thing to nearby burnable thing. Fire is strongly responsive to wind, and a tiny fire can be encouraged by human breath. Fire will live in one place if it is periodically given new things to consume. If constantly attended it can be kept alive indefinitely, neither raging nor dying. It crackles and spits, depending on what is fed to it, and it has a characteristic odor that can be detected very far off downwind. It makes choking clouds by day and a glorious light at night that pushes back darkness. Water fights fire, and a heavy rain can kill it. Tossing a bark bucketful of water on a fire makes a big quarrel with a scary hissing and a white cloud. In times of damp and cold to be just the right distance from a fire is to be as happy as in golden sunlight; but to be too close or let it touch you is terrible, because fire kills and eats living things. Left alone, fire may dwindle into dark red lumps, increasingly frosted with grey-white ash, but up to the time of its death it can be brought to life and made as vigorous as ever by feeding it bits of fire-willing grasses and twigs, and blowing softly upon it.

So the lore may have accumulated, growing into a body of technology exclusively man's. It grew from close observation and continuous experiment, for few things have been as continuously studied by man as fire. In ways that we can now sense only dimly, fire engaged man's spirit as well as his mind, perhaps beginning his preoccupation with haunting thoughts of time and destiny.

A particular conjecture may be ventured. In the ancient days before fire was tamed, three children were trapped late one summer afternoon by a brushfire speeding through a small valley. The smoky, wind-fanned flames forced them to take refuge in a fen-fringed pond. They were thirteen or fourteen, almost ready for admission to the adult company of their band; they were two males and a female. Like most mammals they were nimble natural swimmers, and after waiting alertly until the uneven line of flames had swept past, crackling and smoking, they felt an exhilarated release from fear, and boisterous play began and laughter rang in the smoky air as they splashed and ducked each other.

When the danger was clearly past and dusk was darkening the valley, the children climbed out to rejoin their tribal band beyond the ridge. On their way they came upon a fallen tree, still burning where a dead branch joined the trunk. Drawing courage from each other, they poked at the fire with long sticks, prodding the flames. Dusk had almost deepened to dark when they broke off to climb the valley slope, winding along up the slope in file, each child excitedly carrying a flaming torch back to the band.

FIRE IN THE CAVE

SOMEHOW the magnificent feat was done. We cannot know how the technical triumph was managed; all we know with any confidence is that it was done a very long time ago. The set of our minds today invites us to imagine a single inspired Fire Tamer, a creature of such intelligence and courage as to leave all man in his debt. But it is much more likely that fire was brought under control by hundreds of individuals who learned, and lost, and rediscovered the subtechnologies over uncounted years. Fire became a strange live thing that conferred priceless advantages upon the bands of men that possessed it. They taught themselves to feed and shelter it, to carry it in torches and in smoldering coals and punk, to protect it from wind and rain. In return they could employ it for light and warmth, for hunting and cooking, and to frighten away terrible predators. What they did not realize was that this powerful possession directed their species on a new course, sending it down an evolutionary express lane.

Although details of fire's domestication are unknown, we have considerable information about the shadowy men who achieved it. *Homo erectus* appeared about 1.3 million years ago, and he was a marked improvement on his Australopithecine predecessors. He was, to begin with, no horror-movie hairy ape. He walked upright, and was within our height range, about five and a half feet tall. His brain was significantly bigger than that of his predecessors, and also approached modern man's, with a cranial volume of 750 to 1,400 cubic centimeters, compared with our range of 1,000 to 2,000 cc. There are no data whatever about the color of his skin nor the presence or absence of an epicanthic fold. Little about his body would catch our attention today if we should chance to encounter him in a locker room. Except for his skull, his skeleton was so much like ours that it would take professional measurement to note that the long bones were thicker-walled. After a glance at his features, though, we would be tempted to move to another locker-room bench. The primitiveness of feature would disturb us: a large but slightly receding muzzle, and a low-sloping forehead with formidable brow ridges. It was a child's drawing of a face, pre-barbaric but nevertheless hauntingly human.

A skillful toolmaker, he was able to fashion flint and similar rocks into impacting, cutting, and scraping instruments much superior to hands and teeth for specific tasks. Curiously, his stone tools did not improve continuously in design and workmanship; they simply attained a level of adequacy and held there, or nearly there, for hundreds of thousands of years. At first this puzzled some archeologists, encouraging them in the opinion that Erectus must have been a slow learner if not a dullard, though it is difficult to understand how such a creature could have managed the intellectual feat of domesticating fire. The view now is that it may be the anthropologists who have been the slow learners, misled by the extent to which modern man is obsessed by tools, projecting on them an extra freight of aesthetic attributes. Early man may have regarded his tools dispassionately, something to be made as needed and discarded after use. They may have been his disposables, his beer cans and cigarette packs, not objects on which to lavish attention and carry about as prized possessions. Another theory is that he may have regarded his tools as semi-animate creatures and thus not appropriate for change from their sacred prototypes. Many clues suggest that primitive man was an exceptionally conservative fellow. Whatever his views, the life he led tended to stabilize tool design. He was generally too anxious and harried to develop the leisurely craftsmanship needed to work stone into sophisticated forms.

He was hungry and omnivorous, eating such animals, fish, birds, vegetables, nuts, fruit, and roots as his skill and luck could supply and his stomach stand. We don't know how much speech he had—sound leaves no fossils—but it was probably not much. It must have been more than the rudimentary grunts and howls that other primates employed, and less than a structured communication system able to capture ideas and share experience. We don't know if he laughed or sang.

He was a singularly restless creature, far more so than his African predecessors, moving on to new lands that Australopithecus never saw. Pre-eminently a creature who needed to see the other side of a mountain, he spread over Europe and Asia but not, so far as evidence shows, to the Western Hemisphere at this time. His world was physically more turbulent than ours, with widespread volcanism, the climate unstable for reasons not yet fully understood.* Gigantic ice sheets periodically pushed down from the Poles and out from mountain complexes, tying up so much of the planet's water that the oceans were hundreds of feet lower, uncovering extensive bridges.

Man was then prey, in frequent peril of being stalked, killed, and eaten. He was the victim of bear, wolf, the hairy aurochs, giant hyena, elephant, and the saber-toothed tiger—enemies that variously outmatched him in speed, size, strength, and natural weaponry, and that were his superior in acuity of smell, hearing, and night vision. His numbers—we don't know the populations of early man, but the total number of men on the planet must have been less than live today in a small city of a small country—were also rigidly limited by food supply and disease. He was very far from Rousseau's innocent vision of a noble savage thriving on natural food. Studies in paleopathology suggest that many diseases, including yellow fever, amoebic dysentery, syphilis, yaws, and fatal parasite infestations, were established killers even before Erectus evolved; new diseases such as leprosy and typhoid fever evolved along with him. (He was spared measles, mumps, cholera, and war, the diseases made

* It may well be that our climate is unstable too. Historically recorded man may be like the butterfly that flutters for a few weeks in June, alighting for a moment on the snowplow waiting by the barn.

possible by the high population densities that agriculture was to create.) As a consequence of restlessness, climate change, predators, famine, and disease, the life of this man with the primitive face was short and stressful. In the merciless play of evolution, stress is a forcing function, a wild card that bestows survival for a time on adequately adaptive species. Man's luck in this game has been to draw an enlarging brain, and it has made a pretty experiment.

Of course the experiment has been under way for only about a four-thousandth of the time that life of some sort has existed on the planet, so that continued success is very far from certain. Catastrophic die-offs of species have happened repeatedly in the past, so often as to make them a recurrent theme in the somber music of evolution. The mechanism of these events seems to be not so much an absolute failure of adaptation as a failure to adapt fast enough and extensively enough. The extinct life forms are those that were caught so far down a path of specialization as to be unable to scramble back when their environment fatally changed.

Of all the perils that deviled early man, the only one against which we now seem reasonably secure is danger from large predators. (It is difficult to devise a plausible scenario in which the hairy aurochs gets back his own, although less difficult to imagine other circumstances in which other predators might harry small bands of *Homo sapiens* during the late stages of a die-off.) For all our technical tricks ranging from domesticated fire to thermonuclear fusion, we remain wholly vulnerable to small variations in solar radiation, and variations in stellar radiation are not a rare thing in the observed universe. Catastrophe could come on little cat feet. Disquieting evidence is mounting that the abnormally benign weather of the last 9,000 years is drawing to a close. If our Interglacial is in fact over, the first sign will not be advancing polar ice but a change in equatorial weather patterns. It will be persistent drought in the subtropical latitudes that now precariously support hundreds of millions of people.

Fire directed us toward civilization, and now we are riskily dependent on this structure. We have means for rendering the planet lethally radioactive, and could achieve it by anger or by inadvertence. Our systems for food production and distribution, and for the restriction of war, appear to be both flawed and rigid, seemingly as likely to splinter and collapse under stress as to adapt with sufficient speed. Some sober technical people, projecting the curves that seem to define our civilization, argue that we live in the final years of a trial to see if our institutions can adapt quickly enough to avert catastrophe. Cassandras are no new phenomenon, of course, but not less uncomfortable when stringy-haired seeresses are replaced by men who prefer computers to crystal balls.

¶ IN THE SCALE of hundreds of thousands of years, time is like a great blue mountain range. From one of its peaks there is a magnificent view, a panorama of the circumstance of the creature from whom we are descended, how he lived and learned, shared his learning and died. We can faintly make out distant valleys where he acquired such ideas as home, family, restraint, and duty. There are other places that marked the beginning of rites of propriation and praise. Viewed from the eminence of three-quarters of a million years, we can see that fire was of critical importance in the making of man. In at least seven significant ways fire was a Promethean gift that has led this species to be, at least briefly, lords of the planet and perhaps of the solar system.

In the first place fire made an all-important difference in the struggle with predators. The blazing pine knot snatched from the fire and hurled, shedding sparks, at the eyes in the dark, tipped the balance of terror. This is not simply hypothesized; it is graphically revealed in the layers dissected in the cave at Dragon's Hill at Choukoutien. A cave offers protection to any animal from stormy weather and is a lair for rearing the young; without it or without at least some rocky overhang, it is necessary simply to huddle miserably in the open. Precisely because they were desirable shelters, caves were anciently the home of the most powerful killers. In the oldest strata at Dragon's Hill lie bones of the tiger, cave bear, and giant hyena, along with broken bones of their kill. Atop these are chaotic layers reflecting a long interval of disputed tenancy, as man took possession for a time, and was driven out, and later came back to struggle again. (It is tempting to imagine that these strata represent the time when man was learning to keep and carry fire, the defeats marking tragic interludes when from ignorance or mischance his fires went out.) Finally the eloquent strata reveal only human remains, and the bones of the small deer that were his prey. With the light, smell, and terror of fire at his command, man became king of the cave.

The second of fire's gifts was a corollary: it made possible new and sometimes very productive hunting techniques. A problem for man was that almost all the animals he hunted could outrun him, and were as keen or keener in nose and ears. The lethal range of the thrown club or spear or stone was not large; when firearms were in the last five hundred years introduced to primitive peoples, it was the kill-at-a-distance capability that invariably fascinated them. With torches to set grass or brush burning downwind along a crescent front, a feast of game could be maneuvered into the range of waiting spears and clubs. Many animals hunt in packs but the fire-drive was a different order of accomplishment, requiring premeditation, strategy, and generalship, human counters to superior speed and dangerous natural weaponry. Sites have been found in many parts of the world where thin traces of charcoal indicate that fire was used to drive herds of animals toward death traps, sometimes through funneling canyons and over cliffs, as part of what must have been elaborately organized mass hunts. Perhaps the most remarkable one yet uncovered is at Torralba in Spain, where 400,000 years ago bands of men employed fire to force twenty-ton elephants into boggy ground so that, partly immobilized, they could be killed with spears and stone or bone knives. There were recurrent hunts and feasting here, and the bones of the elephants were systematically cracked with stone tools to extract the prized marrow. The great skulls alone are missing, and no one knows why they were carried off.

Fire's third incomparable gift was warmth. It linked fortunately with man's restlessness, for now he could range over immense new areas of the Earth, carrying his microclimate with him, relatively undeterred by the great epochs of glaciation that swept over much of the planet. At the time of the great hunts in Spain, Torralba was colder than it is today. At Dragon's Hill the climate varied from somewhat colder to a great deal colder, depending on which part of the period of human occupancy is considered. During an Interglacial, the Choukoutien climate was like that of Chicago today; on the slopes of the Second and Third Glacials the weather was harsher, like central Russia or western Canada now. In terms of physics, fire was an advance for man of high importance: for the first time he had access to an energy source other than the Sun's radiation and his own metabolic heat.* The importance went beyond simple comfort in colder environment; it meant that, with fires to warm him, he did not have to consume the same number of calories to stay alive and vigorous. Food is a kind of fuel, as can be observed by comparing the size of meals eaten by

* The increase has been estimated as a doubling. Before fire, his food and the Sun gave man about 2 kilowatt-hours per day of energy. Campfires for warmth and cooking raised it to 4 kwh. In contrast, energy consumption in America in 1970 came to 268 kilowatt-hours per citizen per day.

lumberjacks and office workers. One of the reasons why we do not put away the gigantic meals that our great-grandfathers did is central heating.

Time after sundown was the fourth Promethean gift, of almost incalculable effect. Before fire, darkness disabled men for longer each night than was needed for sleep. The keener low-illumination vision of nocturnal predators meant a heightened risk of being stalked and killed by starlight. Around a blazing campfire, however, conditions were altogether different, providing not just protection against sudden death in the dark and warmth against night chill but also an increase in time for useful activity. The gain obviously varied by latitude, season, and other factors, but it has been estimated that it averaged three hours per day for every man. But the benefit cannot be so simply computed, because the firelit hours were spent in essentially human activity—toolmaking, skinning game, cooking, and above all practicing the unique idea-sharing technique of speech. (We do not always understand imaginatively the staggering implications of speech, possibly because in our daily lives it is the means by which we are most frequently bored; but to an animal staking its chance for survival on intelligence, speech was a communication system of exceptional usefulness, providing an efficient way in which men could guide, support, and teach each other.)

The hours by a fire were, judging by aborigines studied in recent centuries, a favorite time for improving a tribe's weapons. Wooden spears were sharpened by charring the tips in the fire and then scraping off the char, and a spear that had developed a curved shaft could be straightened by judicious warming. This was also a time for shaping spear-throwers, those ingenious notched sticks or bones that, by increasing the effective radius of the arc described by the human arm, increased the velocity of the thrown spear. Stone knives and bone and antler stilettos were born in the fireside armory. Scraping and cutting animal skins by the fire led to better clothing to protect against wind and snow, and for investigation of all the useful things that could be made from leather and fur, thongs and thews. The stimulating effect of the campfire hours on speeding the development of speech must have been tremendous, for here people gathered in a single spot each evening, in comparative security and comfort, sharing warm food, their minds caught by the distinctive way of a fire to encourage thought forward and backward in time: hunts past and planned, strategies for tomorrow, perhaps a move to a new campsite, triumphs and fears, omens and magic, love and laughter and the beginning of song. Fire was the bright, glowing place encircled by known, familiar people, the focus of the band and especially the family, the protected place for the young to grow and the disabled to recover, the core of that deep and powerful concept called home.

Fire's fifth gift, cooking, had like the others a host of widening effects. Lamb's "A Dissertation upon Roast Pig" grates a little on modern readers; it was hardly necessary for Peking Man to burn down a notional barn to discover the special succulence of roast pork, this being an old story from fire-drive hunts. Still, Charles Lamb was accurate in one respect: however accustomed one was to cutting and chewing hunks of raw meat, it was indisputable that cooking made it taste better and easier to chew. But the difference went deeper than taste and convenience. Cooking made food more digestible, with more usable calories per pound, and in the case of meat it made edible a greater proportion of the kill. By partially breaking down the fibers and cellulose that, when raw, were not within the capabilities of human digestive systems, cooking much enlarged the effective food supply and pushed further away the population ceiling of starvation. There were other interactive effects. Most primates have digestive systems that appear to have been originally engineered for a vegetarian diet, which is both bulky and geographically restricting. By the time man was lighting his first campfires, his hunting and cooking skills began to provide him with a food supply that was less place-limited and more compact, richer in the protein and fats that a mobile and restless creature would need.

Because no evidences of fireworthy clay pottery have been found near these earliest charcoal hearths, it has sometimes been assumed that the first cooking was restricted to grilling and roasting. The question is unresolved, for some men have theorized that skulls could have been stewpots, their apertures closed with clay, and this could have been the way by which the firing of clay was discovered. Also many primitive peoples make liquid-tight cooking containers from basketry or leaves, tightly woven and then lined with resin. Such containers are not directly fireworthy but they make serviceable soup cauldrons and stewpots when fire-heated stones are dropped in periodically to keep their contents at a boil.* Stones used in this way undergo thermal shock that can craze their surfaces in a distinctive fashion. Many stones marked in this way have been found, although not in association with the earliest hearths. They are known, not illogically, as potboilers.

The sixth legacy of fire was a temporal awareness and constraint that inevitably led man toward planning and organization, and set his course in the direction of civilization. Most animals other than man seem to live chiefly in the present, and such of their actions as reflect the past or foreshadow the future appear to be of an instinctive character. But early man could not have kept and used fire if he had been a creature living in a feckless now. For him fire was not something that could be extemporized. As a heritage of the band, fire was a precious possession that required unbroken attention and care. It had to be sheltered from rain and wind, fed from stocks of collected fuel, watched over day and night by chosen firekeepers of great steadiness and reliability who knew that letting it go out could be a disaster for the

* What was early man to think of the strange thing called boiling? As an observant fellow, he must have noticed the rushing sound as the liquid grew hot, and then the excited turbulence, the white vapor that rose and vanished, and the slow mysterious disappearance of what was boiled. We cannot know his thoughts, but it is plausible that boiling could have given him concepts of something present but wholly invisible, leading to ideas of spirits that could dwell invisibly in everyday things.

entire tribe. Among men living this way for scores of generations, fire inevitably became the focus of the band. No longer did a hunter feed upon his kill directly at the scene; instead, meat whenever possible was brought back to the campfire, to be cooked and shared with others. Fire thus had the effect of creating a structured way of life, imposing restraint upon individuals who performed specialized duties for the group. Its cohesive effect encouraged the development of individual occupations such as wood-gatherer, firekeeper, hunter, and cook.

Although fire was in these basic ways a civilizing influence, man was—and is—resistant to rapid movement in the direction of learning to live benignly with his own kind. Some of the bones dug up at Dragon's Hill are troubling to our eyes. They are human bones, partly burned, cracked skillfully in the longitudinal fashion that signals extraction of the marrow within. Perhaps there was no taboo against cannibalism among these men with the haunted primitive faces. Perhaps the startling bones came from some barbaric rite of sharing the courage of a dead leader. Or possibly they indicate nothing more than desperation during some winter of cruel famine.

The final Promethean gift came when fire reshaped the human face. After hundreds of millennia, cooking had altered the form of the skull. It happened because the great bony brow ridges, which had been the upper anchors of muscles needed to rip and grind raw food, became less functional. Over thousands of generations those massive protuberances gradually dwindled, and the heavy-walled bone from which they arose could grow thinner, and foreheads could grow larger and more nearly vertical to house the wondering, speaking, fire-building brain within.

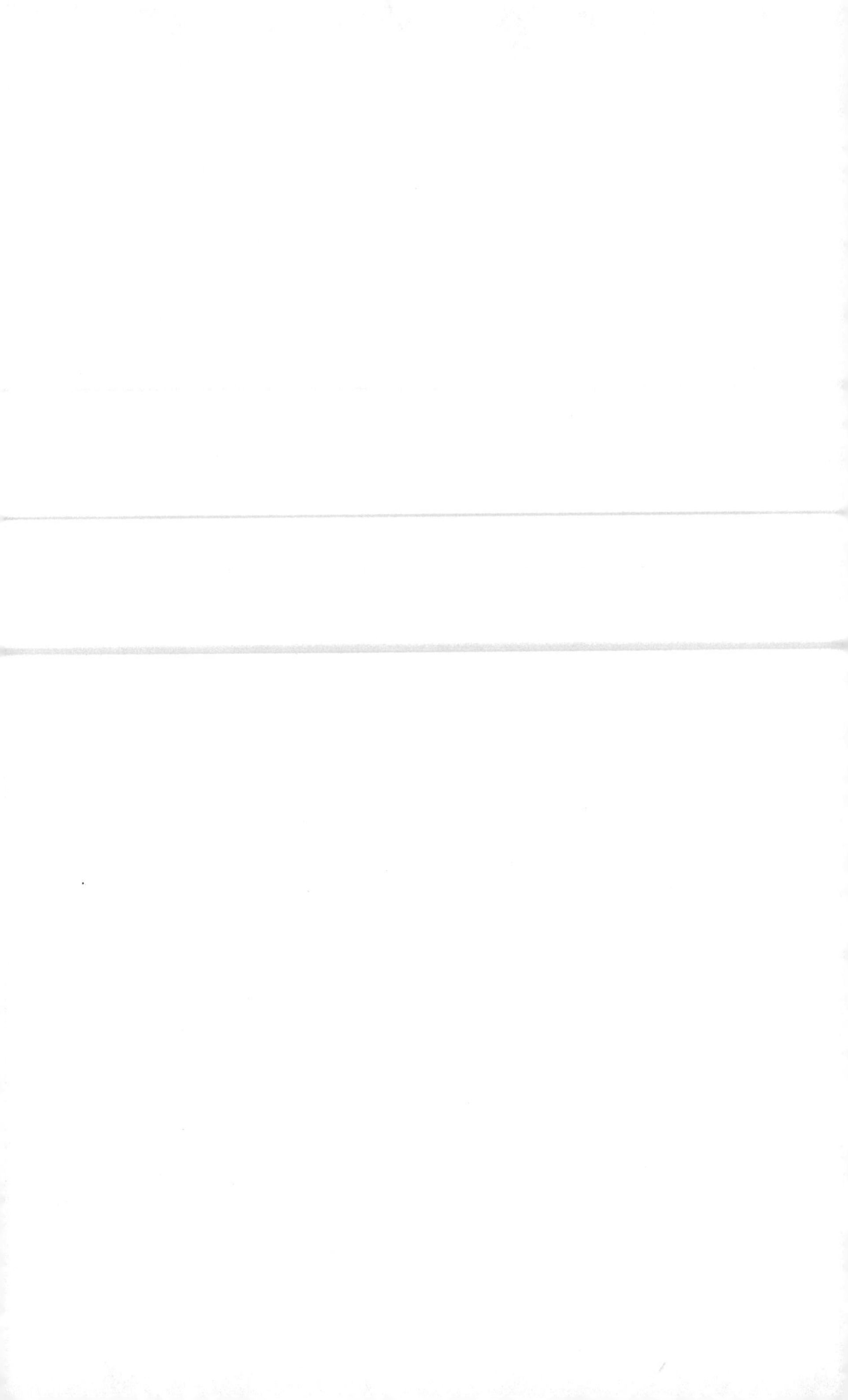

CREATING NEW FIRE

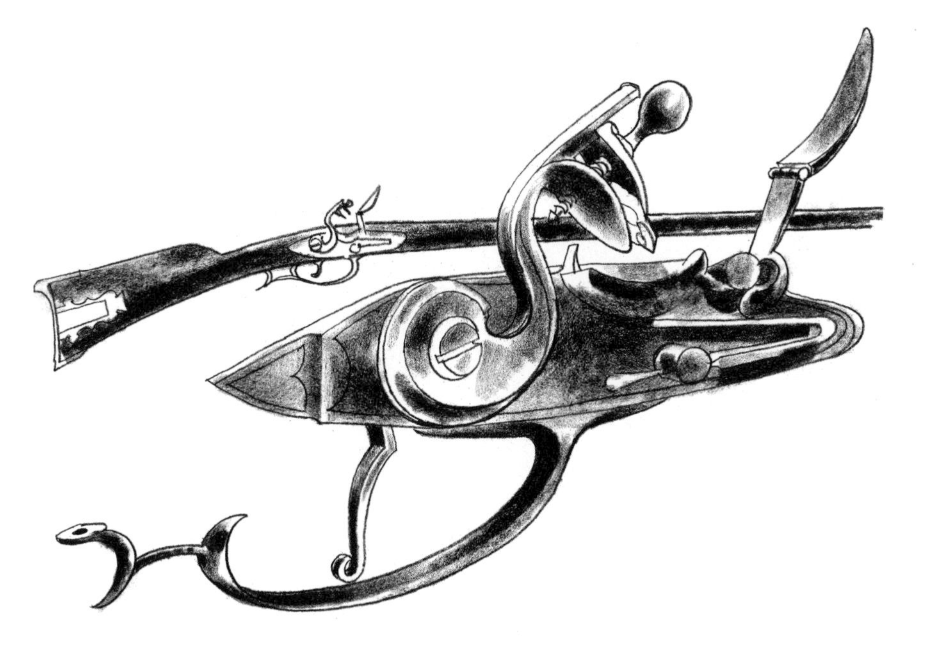

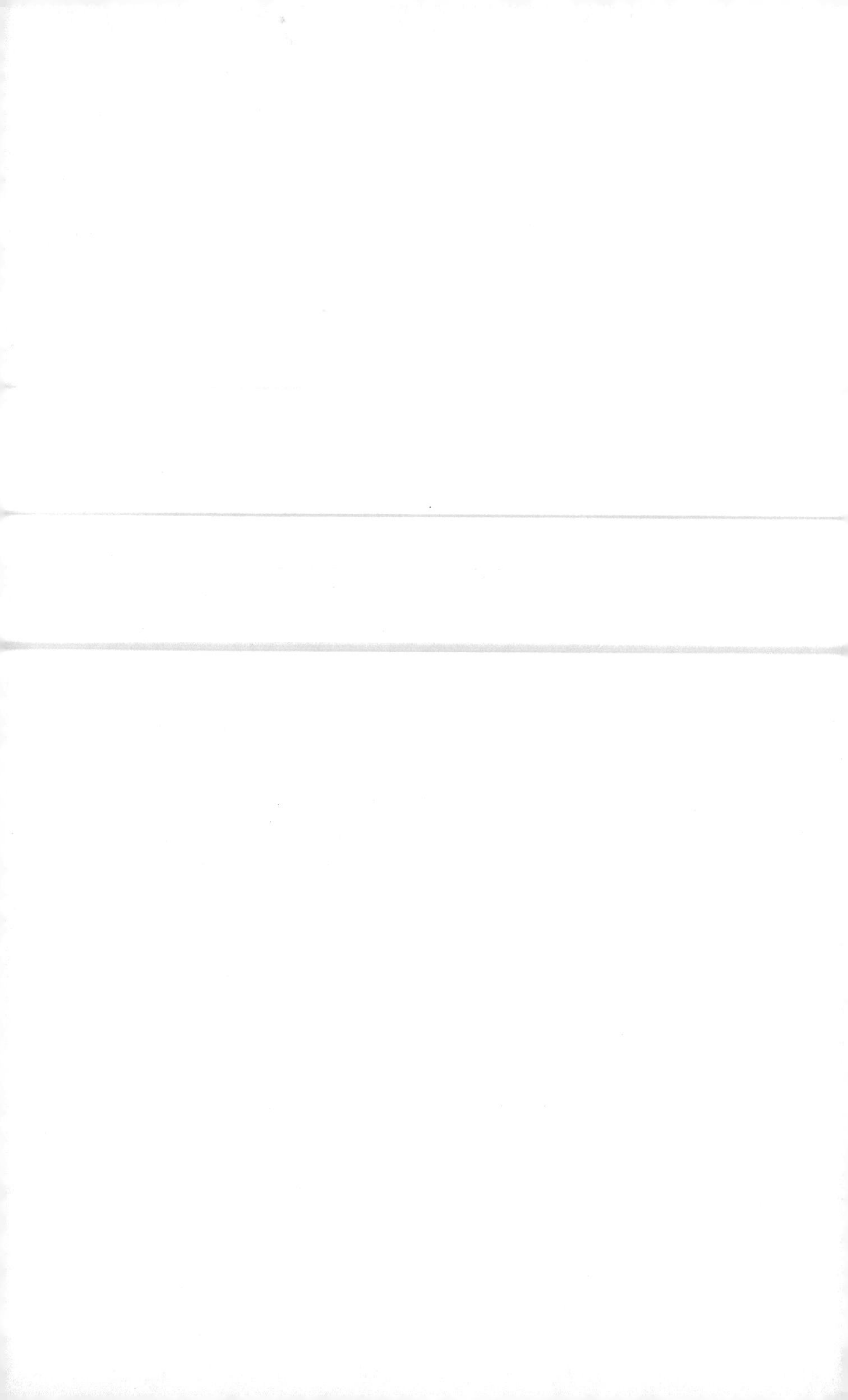

LIGHTING a fire is so ridiculously easy now that the special tools and skills which were needed for thousands of years have been almost wholly forgotten. On the mantel of a house in the country, stored in a trinket box from Hong Kong, there may well be folders of paper matches bearing advertisements for, say, a gourmet diner in Lima, Ohio, or a recommendation that you better your circumstance by learning refrigerator repair, or offering printed testimony to the togetherness of Sally and Herb. Quite possibly there is a cardboard cylinder containing long fire-lighting matches, obtained by Cousin Madge from a shopkeeper who also proffered color-inducing fireplace chemicals, ornamental tiles for every occasion, carved seagulls and bears, improving aphorisms lettered on varnished tree sections, and disguised kerosene pots to help kindle a homey blaze in the absence of any skill whatever. Gift shops and suburban hardware stores abound in fire-making assistance, including electrical and chemical aids for the ignition of charcoal, the latter employed by those who are untroubled by a petroleum-flavored char on their backyard protein. It is difficult to account for this profusion of knickknackery except on the assumption that a substantial market exists for devices that offer facile solutions to nonproblems. In shabby loft buildings in New York and San Francisco, in Kowloon and Taiwan, sit wonderfully pragmatic entrepreneurs, shrewdly considering what was big last year and what might hit this one. There are the tycoons of bric-a-brac, skillful at exploring the ramifications extending from usefulness to convenience and from convenience to a kind of rhinestone-studded silliness, which can sometimes be El Dorado. Ingenuity in mechanism spans a spectrum from, say, a cheap electric clock, where a high level of cunning can be seen in the economy of its interior stampings, to an electric tea cozy, where the cleverness lies almost wholly in conceiving a difficulty that mechanism may avert.

The mantel of a house in the country may also hold a stock of wooden kitchen matches. Devised in a time when there was wood beyond imagining, American in their large, heedless practicality (in contrast to fussy little wax matches of the Mediterranean countries), cheerfully incautious in their willingness to combust on slight provocation, these kitchen matches are presently disappearing from everyday life, like milk bottles, flypaper, and beer-can openers. For generations they lived in a wall dispenser near the coal-wood stove. They could be struck (leaving a dark wake) on any surface except damp wood, coming alight with a noxious chemical sneeze likely to taint any pipe or cigarette lighted during their reactive interlude, as was usually necessary in brisk wind. Their agreeably easy ignition could also be (at the least) a nuisance when a handful of them in a tight pants pocket could catch fire, producing outraged thigh-slapping and the immediate dancing of a jig, with hoarse cries. Their peripheral uses included digging out pipe dottle, or being whittled on the spot into a toothpick, a filler for an enlarged screwhole, or an applicator for a dabbet of glue or a drop of oil. Uncounted small boys found that an assemblage of match-heads made a small, highly untrustworthy backyard explosive, and the left-over matchsticks could be used as small stockades or corrals for the penning of interesting beetles. It was widely known that a kitchen match, put tail first into the barrel of a cocked BB gun, would sometimes ignite on impact if discharged at a rock near by, thus producing a pretend fire arrow or an incendiary bomb, depending on the war being simulated. The uses of the kitchen match extended to demonstrating one's lifestyle by the way in which it was lit: drawn upward along the shank, for the sportsman; flicked with edge of the thumbnail, for the sophisticate (a mode enlivened by awareness that sometimes a bit of burning phosphorus could lodge excruciatingly under the nail); or even struck horizontally along one's front teeth, a fierce and startling gesture felt likely to convince onlookers of one's unexampled toughness.

There is also a possibility that a cigarette lighter is at hand somewhere, although a lesser chance that it will work if not in regular use. Lighters, like dogs, are mirrors of their owners, and we are moving toward the folk belief that you can trust a man whose lighter reliably leaps into flame. If it doesn't, he may simply be careless about flint or fuel supply, but there is a real chance that he may belong to the company of people who live on defeat by mechanism. These are the souls who are nourished by small incompetences, cherishing their ineptitudes as a badge of merit, gaining assurance from demonstrated incapacity. Their taps drip, their bathroom cabinets burst with Band-Aids, their car batteries stand ready to expire. These are the ones among us who display their defeats as though they were mileposts on a long march toward total, or thirty-second degree, incompetence.*

Most cigarette lighters have a charming comprehensibility about them. Look in the next country store you visit for a flyblown card on the shelf near the pipe tobacco and you will probably find a cheap lighter, a Japanese copy of a Zippo, for ninety-eight cents. It will not be particularly well made, with its lid hinge a trifle askew, but it will strike fire with reliability so long as you keep it in benzine and in the little cylinders of cerium alloy that masquerade as flint nowadays. It is something that would have kindled immediate envy from earlier Americans, especially pioneers, trappers, and mountain men. Quickly grasping its workings, they

* It is tempting to ascribe the peculiarity to early environment. A magazine editor I knew was raised in rural Maine and thus could be assumed to be natively at home with objects, but if so he had dwelt too long in Manhattan. One morning his car began to make atrocious and costly noises, so alarming him that he drove immediately to a repair garage. There a mechanic gravely removed a small tree branch from beneath the car, charging him $12.50 plus tax. When I asked if he felt discomfited about this, my friend confessed that, on the contrary, he was so relieved that the engine didn't have to be replaced that he had tipped the mechanic two dollars in a fit of exhilaration.

would cheerfully have traded a bundle of prime beaver pelts, a firkin of whiskey, or possibly an extra woman for so fast and easy a firemaker, indifferent to wind and damp, so speedy a means to start a fire to brew tea, cook food, and warm an achingly cold body.

¶ CREATING FIRE requires energy. Something that can burn must be brought up to the temperature necessary to sustain the continuing reaction that is fire. Today we can tap many forms of energy for ignition: friction or shock, forms of mechanical energy now often used in many cigarette lighters to ignite intermediate flames of benzine vapor or butane gas; electrical energy, expressed as a spark discharge or by resistive heating of a wire; solar energy, focused by mirror or lens; and chemical energy in a variety of forms that range from the catalytic heating of platinum in a hydrocarbon environment to the use of friction to destabilize potassium chlorate or phosphorous compounds to ignite the tiny torch we call a match. We can now create fire by altogether bizarre means: a thermonuclear weapon can set wood ablaze from a distance of more than twenty-five miles. If your taste is less apocalyptic, it would be technically simple to light a candle through a process triggered by the arrival at Earth of photons that have been traveling for millions of years from an unimaginably distant island universe. For everyday fire everywhere on the planet, the match is now pre-eminent. It is so common that in an economic as well as a philosophic sense we have cheapened the magic act of creating fire, manufacturing more than 500 billion matches a year in the United States alone, at such insignificant unit cost that paper matches can profitably be given away, paid for by their ludicrous advertising messages.

The oldest physical evidence of artificial firemaking yet discovered is far too recent—a mere 15,000 years old—to establish how fire was most often made during the hundreds of millennia after its domestication. This evidence, found in a Belgian site, is a rough ball of iron pyrites, a naturally occurring form of iron, that has been grooved deeply by being struck uncounted times by flint. Considering primitive man's practice of keeping fires alight, the depth of the groove suggests that it may have been a venerated object, handed down as a treasure over unknown generations. Before it, evidence is lacking, and we are again confined to hypothesis and probability.

One group of clues arises from the physical properties of minerals. Sparks can be generated by the impact of flint on itself, by the impact of two pieces of iron pyrites, chert, quartz, and quartzite, as well as by combinations of these rocks. None of these combinations is as fruitful of sparks as the flint/iron impact, but they are at least physically possible, and prehistoric men could be, as we have seen, uncommonly clever fellows. On the other hand, aboriginal peoples studied in recent centuries have generally (though not solely) made their fire by friction of wood. Up to the advent of the Iron Age, the firedrill may have been the more common method almost everywhere; after iron grew commonplace, the flint/iron method dominated in the civilized world and was used by the Greek and Roman civilizations.*

No wooden firemaking tools have survived from ancient times, nor could they be expected to. Wood mostly vanishes in a handful of centuries, unlike stone, which has relatively so great a resistance to time that a high percentage of all the stone tools that man has ever made still lie somewhere on the planet. But though the firedrills are gone, a great deal is known about the technique from study of its continuing use by aboriginal peoples everywhere on the globe. Most commonly it involves rotation of a vertical spindle pressing on a fireboard, the spindle being twirled between the palms of the hands in one method, or rotated by a few turns around its shaft of the cord of a bow that is sawed back and forth. In the first method the spinning firedrill is pressed down by the hands, which as a consequence tend to slide slowly down the spindle. They must be periodically raised, which momentarily stops the spinning and allows the unborn fire to cool. The bow-drill avoids this drawback, replacing it with another: downward pressure on the spindle is obtained from above by some hard object, a shell or an indented stone, that adds nonproductive friction to the task and so makes the job more difficult.

* From *The Aeneid:* ". . . quaerit pars semina flammae/Abstrusa in venis silicis" is the kind of line that does nothing for the relationship between harried men and poets. No one having trouble getting a light is likely to feel encouraged when told that he is seeking the seeds of flame hidden in the veins of flint.

Much artful skill is needed for success in this apparently simple process. The wood used for the spindle needs to be harder, but only a little harder, than the wood used for the fireboard. The end of the spindle wants to be rounded to a radius to match a depression gouged in the fireboard. A notch needs to be cut from the edge of the depression to the edge of the fireboard, to afford escape first of the powdered brownish wood dust that initially appears at the point of friction, and then the tiny glowing nubbin of fire that is the goal of all this effort. A pile of tinder must be at hand, and like the firedrill parts it must be truly dry, free of the least trace of discouraging dampness. Hard physical work is required to nurse fire out of wood friction. The twirling method is such vigorous and pace-sensitive exercise that among some tribes it is a two-man task, each man taking over as the other's pace slackens.

The bow-turned firedrill isn't as demanding but it is work enough and takes longer. The effort cannot be slackened as smoke arises from the friction point; it must be sustained and if possible increased, for the little glowing kernel appears only after dense smoke is pouring up. The moment a bit of fire appears the tempo changes, but the task is scarcely less difficult. (To get an idea of the problem, envision snipping off a tiny tip from the end of a glowing cigarette and then quickly, deftly, reliably converting it into a blazing fire.) There are countless natural tinders: dried cattails, shavings of dry wood, tissue-thin layers of bark (cedar is very willing and agreeable), petals from dry pine cones, a mouse's nest, Sun-dried needles, leaves, and grasses, last year's weed tops, all compacted and combined to make a bed for the glowing speck of fire. At the beginning the speck must be breathed upon with great delicacy, for this is a time when a little bit of blowing is too much, and just a little less is not enough.

A different method of creating fire by wood friction has been observed among some Australian aborigines. They make a small split at the end of a piece of wood, introduce a pinch of Sun-dried kangaroo dung in the crack, and then saw rapidly on the crack with a wooden spear-thrower, a process that, given the right technique and dried kangaroo dung, can create fire in as little as 20 seconds of hard work. If we can judge by the variety of methods used in primitive cultures today, it seems reasonable to assume that ancient man made his fire by an equal variety of methods, each adapted to circumstance of place, and each implying acute perception of the locally possible. For their solemn feast of the Sun every June, the Inca imaginatively made "new" fire by ritually focusing sunlight reflected from a polished hemispherical bowl onto a wisp of carded cotton, and it was counted a sad omen if cloudy weather on the appointed day compelled the generation of the new fire by wood friction. Peoples as physically separated as the Eskimo of Greenland and the Yaghans of Tierra del Fuego—each living in damp, cloudy, tree-sparse lands—made their fires by impact of flint upon iron pyrites. No aboriginal peoples discovered anywhere in the world in recent centuries have been totally ignorant of fire at first contact with civilized man; but some groups, including Tasmanian, Pygmy, and Andaman Island tribes, may not have known how to kindle it, living as they did (and do) in wet country where finding and keeping usable tinder was almost impossible.

The Indians encountered by the first Europeans in North America generally used firedrills, both palm-twirled and bow-turned, but the method was touchy enough in damp weather so that they also carried fire with them on their travels. The techniques for portable fire were often ingenious, with a sophisticated practicality not always associated with primitive peoples. The Penobscots, a New England tribe interpreted by the first settlers to be formidable savages, carried fire within pairs of large clamshells lined with insulating Sun-baked clay and ventilated by small drilled airholes. When a trip was begun, a long, thin strip of dried fungus was coiled in a flat spiral, its end lighted from the campfire, and the coil placed within the insulated clamshells, which were tied tightly together with thongs. One coil could last a day, quietly smoldering like punk; and extra coils for longer trips could also be carried. When a campfire was to be made, a Penobscot simply untied the shells, placed a bit of shredded cedar bark against the glowing end, and by careful blowing produced a flaming cedar firematch. *Chiquoqusoqu* was the Penobscot word for the material that made this feasible, a creamy, brownish fungus that grows preferentially in cracks on living rock maple or golden birch. No other local fungus works as well; and even this one requires skilled handling, harvested at the right stage, cut into strips of the correct width, Sun-dried to a point that permitted coiling without breaking or crumbling, either of which would make the whole effort worthless. It is exhausting to think of the amount of observation and trial that must have been invested in perfecting this single technique, the number of alternative materials that must have been tried and rejected, the number of other means of carrying and using burning punk that must have preceded the use of perforated, insulated clamshells.

Other technologies used by prehistoric men also demonstrate—by their subtle observation, delicacy, and implied body of experimentations—that their inventors were very far from dim brutes who passed their days in bashing each other with clubs and dragging females about by the hair. Many of the perplexities we feel about departed peoples actually derive from our own assumptions. The Inca, for one, are sometimes treated with surprise because, despite their preoccupation with creating massive, meticulously fitted structures of quarried rock, they apparently never bothered to invent the wheel. Yet although it offends our wheel-loving culture to concede it, the Inca may have known about the wheel but simply not thought much of something so unsuited to their rough and craggy country. (They almost certainly knew, as did the Egyptians, that rollers were fine devices for moving heavy loads.) So if, misinterpreting their uninterest in the wheel, we incline to dismiss them as primitives, we should instead remember the wonder of those polished hemispheric bowls, surely a magnificent way to create ritual fire under the high Andean Sun.

¶ TOUCH A BIT of iron or steel (the difference being in this case immaterial) to a turning grindstone and it immediately becomes evident why flint-and-steel won over the firedrill as soon as iron became generally accessible around 2,000 B.C. Iron is pyrophoric: when it is struck by flint—an exceptionally hard silica—particles of iron are heated by friction and shock to incandescence, torn free, and sent flying. For a short interval, on the order of a second or more, the particles can communicate fire. Unlike wood friction, no extended period of vigorous energy expenditure is needed; with flint-and-steel the process begins at a point achieved only after exhausting effort with a firedrill. This was a merit quickly grasped by the users of firedrills, which were generally discarded as soon as iron objects appeared in trade channels. Technique was no problem, for the methods were identical once a spark had been created. Many native peoples, including the North American Indians among many others, had ample opportunity to learn all about flint-and-steel from the flintlock guns used to encourage them to vacate their ancestral lands.

Flint is a strange dark stone, found all over the world, formed within limestone in ways not fully understood, of a character that is at the same time both agreeably workable and wickedly mean. It is a stone closely associated with man, being uniquely adapted to manlike purposes. Until metals came into use four to five millennia ago, flint was exceptionally important in neolithic cultures for knives, axes, spear points, arrow points, grindstones, and (with flint flakes set into bone or wood carriers) scythes, sickles, and threshing tools. It was so much prized and so extensively used that neolithic man mined it in a remarkably organized way. Hundreds of ancient flint mines have been found in England and Europe where prehistoric men dug shafts as much as fifty feet deep, using the shoulderblades of oxen for spades, and then dug horizontal crawlways to extricate flint nodules. These brave miners used pieces of deer antler for picks, and lighted their perilous labors with lamps burning animal fat, which left sooty marks on tunnel roofs that can still be seen today.

Even after bronze and then steel replaced flint in axes and weapons, this curious dark stone—which Germans call *Feuerstein*—remained the chief source of new fire for civilized man right up until the 19th Century. Its fateful, not to say fatal, uses were by no means ended when stone spear points became obsolete. For a century after the gun was invented it remained a limited contrivance, because discharging it required a glowing slow-match or red-hot poker brought along for the purpose. It was only after the wheel-lock and flintlock appeared about A.D. 1500 that the gun became, in portability and readiness for immediate use, a practical weapon for everyday death. The flintlock was in principle a simple firemaking machine, with a pinch of priming to replace the tinder and conduct fire through the touch-hole to the main charge. Although basically simple, it grew to be a mechanically elegant device, rich in the springs, pivots, cams, pans, pan covers, and all the polished metal ingenuities that gifted gunsmiths could create. It had an exceptionally long life as technologies go, remaining in broad use for three and a half centuries, and serving as the basic infantry weapon in the Revolutionary, Napoleonic, and Crimean wars. In fact it still hasn't become totally obsolete, for many thousands of gunflints are still shipped every year to outback regions of the Earth for trade with flintlock owners. The continued use of these weapons by remote people is not so much antiprogressivism as it is practicality: a flintlock is a do-it-yourself gun, not dependent on a supply of commercially manufactured cartridges.

Flint-and-steel made household fire for at least four thousand years. The task was done this way: A piece of iron or steel was held in one hand just above some exposed tinder, and it was struck a brisk glancing downward blow with a piece of flint in the other hand. When a spark falling on the tinder began to smolder, it was blown upon lightly to encourage and brighten it; then the fire was transferred to and brought to flame with a spill of paper, a wood shaving, or splint of dry wood that had been prepared by being dipped in molten sulphur. (The Romans are credited with the innovation of sulphur splints; they worked fine for the smolder-to-flame transition but smelled terrible and weren't cheap.) Indoors, with deftness and dry materials, the firemaking process could be completed in less than a minute. But it presented problems to the old or dim-sighted, and could be exasperating to the impatient or clumsy. Charles Dickens was fascinated by a study in 1832 indicating that the English housemaid could need from three to thirty minutes to create a flame, even with practiced adjuration and imprecation. Striking a light was enough of a nuisance so that any existing fire which had survived overnight in lamp, stove, or fireplace was always welcome, and it could readily be carried from room to room with a shielded candle. In striking new fire it was desirable to have a piece of flint that was eager to produce sparks —not all flints were equally willing—but it was absolutely essential to have dry and co-operative tinder. If by mischance dampness or even high humidity had affected the tinder, one went next door to borrow fire.

By far the commonest tinder was old cloth, something worn-out from the ragbag and torn into small squares. Old cambric linen handkerchiefs, when available, were highly regarded. Cloth tinder was charred before first use to make it particularly hospitable to sparks, and kept except in use in a tightly closed drawer or tin. Many natural tinders were also used—Sun-dried layers of bark, leaves, and grasses; thistledown, milkweed, and dandelion balls; and the floss from the cocoon of the silkworm. City dwellers who could afford it might buy their tinder from street hawkers selling commercial preparations known in various times and places as touchwood, German tinder, and amadou (the latter from the French verb *amadouer*, to coax or attract with bait). These were often made from the fungi that grew on oak, beech, and birch, pulverized, boiled, dried, and sometimes impregnated with saltpeter. It is noteworthy that these fungi were used by (among others) Stone Age Englishmen, American Indians, and 18th Century householders of Paris, London, and Philadelphia. Tree fungi are by themselves rather soggy objects, and their affinity for fire grows evident only after they are ground and thoroughly dried. For so many people separated in time and place to have independently made this modest but highly useful discovery suggests that man once examined his close-in physical world with an acuteness of observation and willingness to experiment that may be growing uncommon today.

Flint-and-steel—the combination often being called a strike-a-light—could be carried loose in a pocket, but tinder required special protection. It could be guarded against dampness by being folded within oilskin or kidskin, or (common in mediaeval times) carried within a tightly stoppered oxhorn. The nobility and the rich were fond of pocket tinderboxes of great elegance, with elaborately carved and jeweled designs. Some tinderboxes presented to the great displayed a kind of Nieman-Marcus lavishness of wonder and ingenuity. In the first order of elaboration, the object was made to appear as something other than itself, a jeweled and lacquered egg, a carved figure or miniature animal. In the second order of elaboration it was not only disguised but dynamic: it did something. Historians of technology think it possible that the idea of flintlock firearms may have been suggested by certain marvelous spring-driven tinderboxes brought from Japan by Portuguese traders in the 15th Century. Somewhat later, members of the nobility and the rich could amuse themselves with miniature pistols that flashed a pinch of gunpowder to ignite a sulphur splint. Later still there were clockwork contrivances to light a candle at a preset time*—analogues of the contemporary time-switch supposed to serve as an amulet against burglars. Such costly and not very practical gadgetry was not for everyone. In most households at the beginning of the last century the strike-a-light, tinder, and splint or shaving were stored prosaically in a box or tin on the mantel or by the stove.

The world of the strike-a-light has been swept away, or almost so, by the friction match. It lingers a little, in principle, in cigarette lighters and in the igniter that an oxyacetylene welder routinely carries in his left hip pocket. It lingered up through 1969 in the Missal of the Roman Catholic Church, where it had been specified that new fire for the Easter vigil was to be "struck from flint." Then in 1970 the requirement was dropped, for reasons that are not clear. Possibly some ecclesiastical bureaucrat concluded that since paschal fire was being created in churches all over the world by cigarette lighters, and since the flint in lighters wasn't true flint, the

* Fire and time have curious linkages in the human mind. One highly popular device in the past was the Sun-fired noon gun, a miniature brass cannon surmounted with a burning glass so arranged that the noon Sun ignited powder at the touch-hole. While it must have made a joyous bang, it is not easy today to understand the amount of pleasure given by such a device, especially in regular use. Perhaps it was welcomed as a signal of time to begin drinking. Perhaps it was perceived as clever, a kind of witticism in brass. Still, a cloudy day spoiled the fun.

phrase should be omitted. (Lighter flint is now an alloy of iron and cerium, the latter a rare-earth metal so hysterically pyrophoric that it is also used in machine-gun tracers.) Perhaps the Church, unconcerned about technical quibbles, simply recognized that new fire is a concept unrelated to its method of origin.

Yet the ancient past has a way of never quite receding completely. A NASA man I know, not deep in middle age, has recounted a vivid boyhood memory of watching a weary Mexican cowboy use flint-and-steel to light a cigarette on horseback. It was done with dexterity and grace as the pony picked its way along the trail. First the cigarette was rolled and sealed in the conventional cowboy fashion and tucked behind the right ear. From a pocket in the worn leather vest came the tinderbox, a brass rifle cartridge. A bent wire running inside brought up the fluffed end of a bit of cord, the tinder. Also from the vest came a small piece broken from a file, and a thumbnail-sized piece of flint. The steel, reins, and tinder were arranged vertically in the bent fingers of the left hand, and the flint in the other hand fetched the steel a rap or two. After sparks caught in the tinder were breathed into a glow, the cigarette was retrieved from its parking place and lighted. I have thought it marvelously fitting that this sharply remembered scene of casual skill was played before a boy who, a few decades later, would be as much responsible as any man for the extraordinary technical virtuosity that was required to send the first unmanned scientific spacecraft on scouting expeditions for scores of millions of miles from Earth past Venus and Mars.

¶ IF MATCHES had just been invented, we would surely think them wonders, little packets of clever self-igniting torches providing fire and light in an instant, needing no skill, asking no physical exertion, as cheap as a drink of water. We would be proud of their invention, interpreting it as confirmation of free enterprise, native ingenuity, or any other concept felt likely to benefit from being testified to. (Perhaps we would worry a bit about the effect of matches on the young, already dangerously exposed to the softening effects of modern life.) Reporters would seek out the inventor of the little wonders, if he hadn't been trampled flush with the ground in the corporate struggle to wrench away control of the new product, and inquire about his working habits, favorite food, prescription for wealth, and his views on politics and the rôle of women. None of these exercises is possible because matches in fact had scores of inventors, spread over several centuries in several countries; and because a reasonably practical product, although announced by promoters as regularly as the sound from an automatic foghorn on a stormy night, was not achieved until many more decades of painful development. Some of man's achievements have been managed with style and grace, an arrow flying to the mark. Matches, in contrast, sputtered noxiously into being over a long period, and it would not be easy to erect bronze statues to creators of products named The Pocket Luminary, Chinese Lights, The Phosphoric Candle and Ethereal Match, Electropneumatic Fire Producers, Lucifers, Vesuvians, The Instantaneous Box Light, Oxymuriated Matches, Bengals, Wire Fixed Stars, Prussian War Fusees, and Locofocos.

The technical development of matches began, in a way, with Paracelsus, a gifted but difficult and notably short-tempered Swiss physician whose writings in 1526 are the first to mention phosphorus, the poisonous, evil-smelling fifteenth element, capable of glowing in the dark and catching fire spontaneously in air. Over the span of many centuries it is difficult to be certain what Paracelsus meant in writings that combined brilliant technical insight, alchemical dogma, and obscure mystical religiosity. He was a founder of pharmaceutics, an advocate of chemical as opposed to vegetable remedies, the first physician to recognize the concepts of metabolic and occupational diseases. At the same time he was a cranky man, touchy and arrogant, with a gift for enmity. He named himself Paracelsus, meaning "better than Celsus," a celebrated Roman medical authority of the 1st Century; his natural name was Philippus Aurelius Theophrastus Bombastus von Hohenheim. He died violently in his forties in 1541, according to his detractors in a drunken debauch but according to widespread contemporary opinion in a fall from a window arranged by a number of physicians and apothecaries who found him unendurable. While it is not proven that he was the first man to isolate phosphorus, it is difficult to see how he could have described it without doing so. Certainly no more likely candidate could be imagined as the first person to have laid eyes on this cranky and toxic element.

Phosphorus was not rediscovered until one hundred twenty eight years had passed.* It was found again by Hennig Brandt, a secretive Hamburg alchemist hot on the trail of a process for converting silver to gold. He made a small amount of the element by a process that began with the residue of evaporated urine, found it of no value for his purposes, sold his secret to Johann Krafft of Dresden, and disappeared from the pages of history. Krafft was delighted with so strange a substance and exhibited samples in Europe and before the Royal Society in London, where the material fascinated Robert Boyle, the brilliant Irish chemist and physicist who was one of the Society's founders. Krafft tried to keep the process for the manufacture of phosphorus a secret, but Boyle, alert and shrewd, was able after several years' effort to make his own phosphorus. In 1680 Boyle and others began selling firemaking devices consisting of pleated phosphorus-treated paper, and sulphur splints. If a splint was pinched in a fold of the paper and pulled briskly out, it would sometimes, astonishingly, burst into

* The time interval casts some light on the solitary and individualistic way by which the great scientific advances of the Renaissance were achieved. In conspicuous contrast is the way in which many modern scientists work in packs, sometimes but not always in organized teams but commonly in a kind of wave-front of collective and sometimes competitive activity. It is hard to conceive today that the premature death of one member of the pack could retard achievement for a century and a quarter. But of course the proposition is negative, and it is hard to be sure about what could have been done but wasn't.

malodorous flame. It was the world's first commercial match. It had a great future but no present, since phosphorus then cost more than gold. After a few sales to the very rich, who seem as a class to have a fitful interest in costly novelties, it was forgotten.

For more than a century flint continued to light the fires and lamps of the world. With the American and French revolutions an interlude of inventiveness began that brought many technical advances, including practical matches. It is conventional wisdom to observe that war advances technology; but in view of the broad strides in science that began near the close of the 18th Century, it seems evident that times of social and political unrest correlate with a general outflowing of creative energies.

By 1775 phosphorus was no longer made with alchemical ritual from urine crystals; it was prepared at greatly reduced cost from animal bones, and the method was published knowledge. At the outset, however, the first approaches to matches were more creative than practical. From France came the Phosphoric Candle and the Ethereal Match, devices in which a strip of paper treated with phosphorus was sealed in a glass tube, and when the glass was broken the paper caught fire. In England there was the Promethean Match, a paper-wrapped glass bead with sulphuric acid inside and a coating of potassium chlorate outside. When the glass was broken (a pair of pliers was provided for the purpose), the paper caught fire. If the pliers were mislaid, daring rakehells broke Prometheans in their teeth.

From Italy came the Pocket Luminary, a small bottle of phosphorus, and when sulphured splints were dabbed within and removed they ignited. In America there was the Instantaneous Light Box, a bottle of sulphuric acid, and its splints were tipped with potassium chlorate (the splints cost four cents apiece, not a negligible price in 1804). From Germany came the Electropneumatic Fire Producer, a staggering piece of technical virtuosity. On demand it created a small jet of hydrogen, generated by the interaction of sulphuric acid on zinc, igniting it in one model by a platinum catalyst and in another by an electrical spark from rosin friction. The inventor of this eerie device was Johan Döbereiner and several of his creations still exist in museums, inducing in onlookers awe as well as wonder about the German cast of mind.

Even after workable friction matches appeared, some inventors appeared unable to bridle their fancy. In 1839 a self-lighting cigar turned up in Austria, adding the flavor of igniting chemicals and charred wood to the taste of tobacco. A later version, evidently designed after complaints about the vile taste, introduced between the match-head and cigar a frilly paper construct resembling the bootees sometimes worn by lamb chops; it was not popular either. Percussive matches made a brief and noisy appearance in the 1850's, resembling cap-pistol ammunition and requiring a plunger mechanism to set the paper strip afire. As late as 1882 the weird motif continued when the Diamond Match Company purchased a patent for something called the Drunkard's

Match. Its splint was chemically treated so that it refused to burn beyond midpoint, and it had a brisk sale for several decades, possibly as a jocose gift.

The main line of development that matches were to follow was friction ignition, and while this was plainly more practical than pocket bottles of acid, the course was not easy. The first friction matches to gain broad use were Samuel Jones's Lucifers, first sold in London in 1829, which were not easy to light and which caught fire with a shower of sparks and a noxious cloud of fumes. Although Jones was sharp enough to use someone else's formula after he observed that it was unpatented, he displayed enough social conscience to print on the Lucifer box:

> If possible, avoid inhaling gas that escapes from the black composition. Persons whose lungs are delicate should by no means use Lucifers.

While most of the early friction matches were cranky to light, a French chemist found a way to replace potassium chlorate in the igniting compound with white phosphorus, and this changed the behavior of friction matches drastically: they became not just willing to leap into flame but in many instances overeager. It was not known at the time, but white phosphorus was a thoroughly bad actor in the chemical family, unstable and dangerously toxic. For almost eighty years match chemists labored to teach it manners, with only partial success, and at a cost in human health and life. What made this foray down a technical dead end sadder was that it was needless. In the 1840's Swedish and German chemists discovered red phosphorus, a compound that was essentially nontoxic and not susceptible to spontaneous combustion. They also had the inspired idea of separating the ignition ingredients and putting part of them on a striking surface, so that a match became chemically complete only when struck on that surface. Such safety matches were in broad use by midcentury. But the strike-anywhere match containing white phosphorus had also grown very popular, being slightly cheaper and felt more convenient at fireplaces, stoves, and lamps. (Some users may have dimly interpreted the safety match as a limitation on freedom.) It was, nonetheless, a singularly nasty product. It could light spontaneously on hot, humid days; rats and mice liked to gnaw on the heads, and could start fires in this way; small children who put white phosphorus matches in their mouths could be poisoned; and as word of its toxic powers got about, it may even have become a handy domestic source of poison for murder and suicide.

The white phosphorus match became a classic 19th Century example of the confrontation between free enterprise and public interest. By the final quarter of the century the manufacture of matches had long outgrown its cottage-industry beginnings to become a large and profitable enterprise, with capital invested in plant, equipment, and raw materials. It is possible that makers of white phosphorus matches believed that chemists were about to make the bad-actor chemical more civilized, and indeed the spontaneous-combustion temperature was somewhat raised in time. But gruesome new developments outweighed these small gains. Workers making white phosphorus matches began to display a disquieting mortality. The problem centered around what factory workers called "phossy jaw," and it took medical experts some time to understand its mechanism: white phosphorus fumes were entering workers' bodies through the avenue of decayed teeth, concentrating in the jaw and creating there a frequently fatal necrosis.

Nearly half a century was required to resolve the conflict between a clearly vicious product and a world-wide and profitable demand for it. Denmark and Sweden began to ban white phosphorus in the 1870's; other industrial countries did not agree to a world convention on the subject until 1905–6, at which it was agreed that the compound would be banned in 1912 or, if banning was legally doubtful, it would be taxed out of existence. Since 1912, strike-anywhere matches have replaced the bad actor with phosphorus sesquisulfide, a compound that is vastly less toxic, not prone to spontaneous combustion, not perilous to factory workers, and unappealing to the hungriest rats and mice.

It seems appropriate that the industry that displayed so resolute a resistance to unduly precipitate action was ultimately unable to preserve the pre-eminence of the strike-anywhere match. Sales of paper, or book, safety matches have, thanks to smokers, stayed close to an all-time high. The strike-anywhere match has in contrast entered a long, sloping decline, as central heating replaced the fireplace, as electric light replaced the lamp, as electric stoves appeared in the kitchen, and as gas stoves, once a glorious consumer of matches, began to be equipped with pilot lights, as newfangled a device as any matchmaker could imagine.

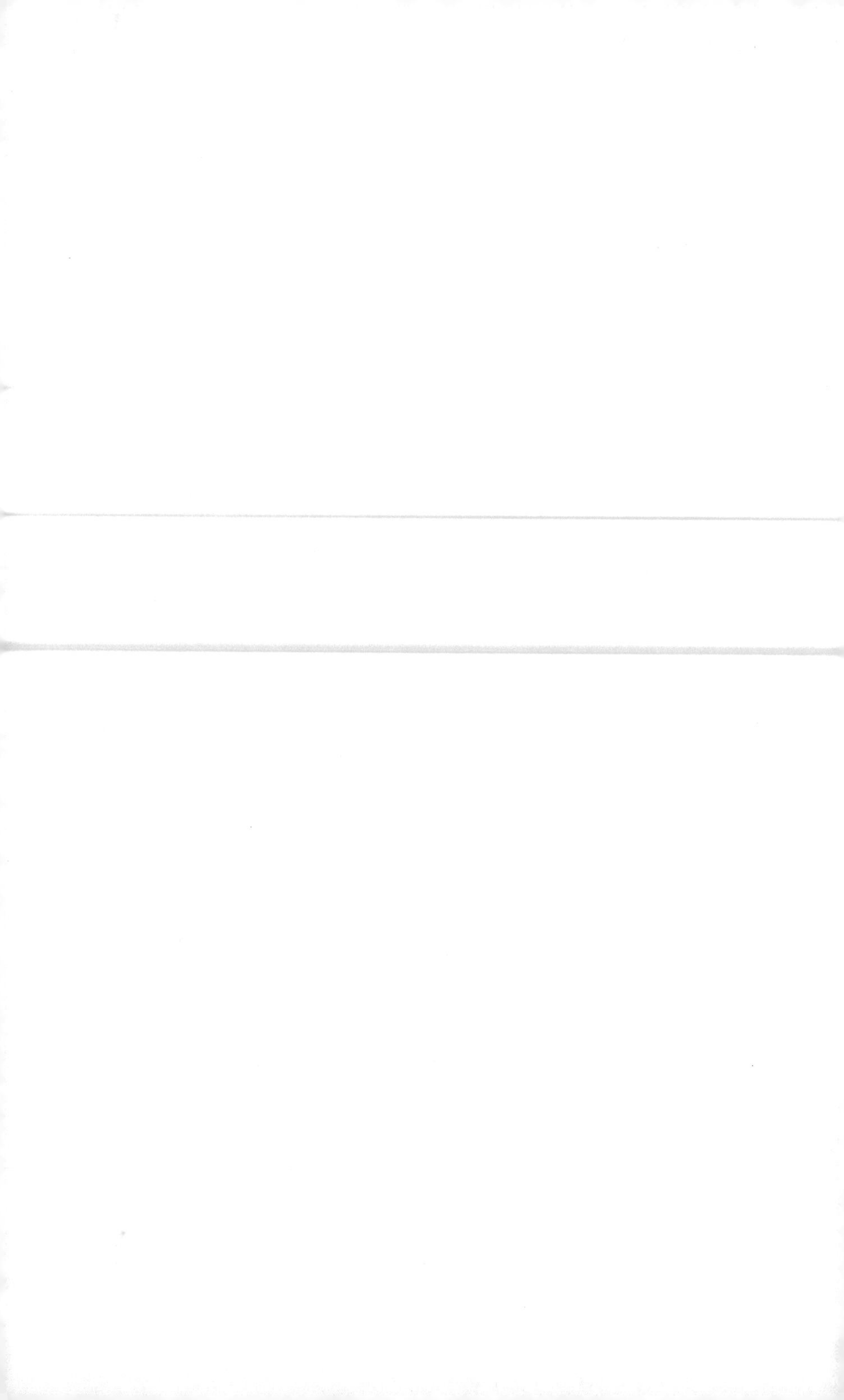

APPLE & BEECH, BIRCH & OAK

WOODCUTTING is full of sensuous and kinesthetic rewards. There is the rhythmic *thock* of the axe, with every second stroke freeing flat-spinning chips of yellowish wood. The special alertness of felling builds up during preliminary notching cuts and rises to a breathless moment when the treetop loses the symmetry of its sway, hovers against the sky for a moment, and then sweeps over with fierce cracking as the trunk thunders to the ground with branches thrashing. When limbs are cut free with diagonal-slicing axe strokes, the ellipses of fresh-cut wood gleam against the bark in the soft forest light. As wood is bucked into carryable lengths, measured multiples of what your fireplace accepts, the saw generates little conical hills of cuttings; coarser than carpentry sawdust, these tiny cubes of pale damp wood creep into boot tops and pockets. When wood is split in the yard by the woodshed there comes a time when a husky billet of hardwood, almost too big and cross-grained to be axe-splittable, nevertheless responds to a precise blow by dropping apart in two even pieces, their inner faces moist and fragrant at this first exposure to light and air. Toward dusk there is an enjoyably tired feeling as the day's yield is stacked for seasoning, the wedge-sectioned pieces fitted evenly, like a carefully built stone wall; only stovewood is allowed to remain in a tumbled heap.

Getting in your own firewood has much to recommend it, if not done of necessity. Felling, limbing, bucking, and splitting are the four basic steps in the process, none difficult to bring off passably and yet none so simple that your technique cannot be bettered. Prodigious exertions are not called for, although it is somewhat more strenuous than golf or other conventional absorbents of leisure. It is work that is likely to blister the hands or crick the back unless undertaken along a gentle slope of habituation. You can do a considerable amount of creative resting in the woods, seated on a log, absently rubbing tired muscles, studying the intricate topography of bark, reading the now-ended calendar of tree rings, smelling the fresh woody fragrances, watching the chipmunk or bluejay come closer to scout the meaning of your silence. Almost alone among outdoor pursuits woodcutting is unaffected by weather, and can be engaged in at all times except drenching cold rain or driving sleet (it is quite magical during a snowstorm). You automatically work at a tempo attuned to the temperature, slowly enough to keep from steaming on a golden day in Indian summer, briskly enough to keep comfortable when north winds icily rake the woods.

Unlike many leisure activities, woodcutting delivers a product in return for time and effort, a pile of beautiful well-split hardwood, abounding in promise of bright fires radiating their magic on winter nights. The degree of satisfaction is curiously high. It is not altogether clear why a filled woodshed should be so reassuring, more so than shelves of cans and jars in the cupboard or a freezer chest chocked with frosted parcels. Some shadowy motivation is buried here, something to do with the peripheral rewards of possessions, and with a need for making experience tangible and visible to others.

Perhaps a clue can be sensed by considering people whose leisure activities do *not* afford a product. If you observe the cameras carried in such profusion by returning travelers as they flow through customs weirs, it is possible to conclude that the profoundest need filled by vacation photography is documentation. Pictures, studied and shown about, function as a kind of certification of experience, and generally they don't bleach as fast as memory. Returning travelers carry, in addition to luggage and cameras, an extraordinary volume of gifts and bric-a-brac. From the balcony above the arrival pens at an international airport, a jetload of tourists tends to look as if most of them were victims of besetting generosity or compulsive connoisseurship. The scene suggests that possessions bought abroad and carried home have high value as confirmation of time well spent, proof less evanescent than airline tags left dangling and customs symbols chalked on luggage. People appear to buy knickknacks for the oddest of motives, not the least of which is the conversation that can be casually constructed about them later. Evidently recollection is for many persons simply not sufficient, and a physical thing or simulacrum image supplies welcome additional testimony.

As for the filled woodshed, it is true that few countrymen whip out a camera to document the achievement of several cords of beech, but it *is* noticeable that conspicuous supplies of firewood are frequently stacked in an open shed in no way concealed from those driving past. Firewood is after all not as inherently invisible as those parcels in the freezer, and a possession that is invisible is somehow slightly diminished.*

Getting in firewood is a many-faceted activity, and not the least of its benefits is a sense of calm. You may stride off to the woods bearing, along with axe and saw, a well-nourished grievance; you are almost certain to return with, if you think of it at all, detached surprise that the matter could have been so troublesome. Woodcutting is a sovereign remedy for a churny mind, a specific for festering concern. It works its spell not so much by substitution—which is what skiing does, it being difficult to cherish a grievance while hurtling downhill in continuous alarm over narrowly averted catastrophe—as by the more subtle method of transference. Using an axe or splitting-sledge is of course a form of sanctioned violence, and it takes only a half-hour's tussle with some

* This peculiar verity was discovered, surely not for the first time, during introduction of air-conditioning for automobiles. Disappointing initial acceptance of this costly extra led some Detroit officials to consider adding conspicuous if unneeded air-intake vents to help a buyer derive a full sense of reward for his purchase. Later it was discovered that the sight of people driving around on sweltering days with windows cranked up did the announcement job, but just to be on the safe side a window decal was also provided. Wearing a sign certifying the ownership of an invisible possession is, when you come to think of it, rather odd behavior.

mulish hickory to mop up any likely supply of hostility. This is nothing so simple as venting one's spleen on poor harmless trees: it is instead a kind of cancellation, perhaps a physiologic process related to the sensations of using the large muscles.

Woodcutting also provides a second form of psychotherapy by presenting a series of small, engrossing, and delightfully soluble problems. A tree is felled exactly where it is supposed to fall, despite a slight lean in another direction; threatened pinching of the saw kerf is avoided by reordering the natural sequence of bucking cuts; an unsplittably gnarled crotch is finessed by relocation of fireplace lengths. It is not necessary to pretend that these are substantial achievements, nor anything more than routine to a woodsman; it is just that the successful solution of random problems, however small, is tonic to the spirits. Certainly it could not have been by chance that Wilhelm II, the last of the German emperors, his restlessly proud and never very stable mind corroded by an awareness of a world war disastrously lost, spent time each exiled day in woodcutting.* But it is not necessary to have the Marne, and Verdun, and the

* It should not be thought that the old man at Doorn laid about him in a welter of chips and sweat; his approach was rigidly decorous. Joachim von Kürenberg, a sympathetic biographer, describes the scene: "The day started with a morning service; the Kaiser read the lesson. Breakfast followed and after that, weather permitting, a walk in the park or wood sawing; old Schulz or Vieke would stand by, holding the master's coat, watching this new occupation with astonishment. They had been in the Kaiser's service for more than thirty years." Von Kürenberg, a guest at Doorn, clearly felt that *Der Allgehöchest* at the woodpile was a monocle-dropping sight.

utter ruin of an empire on your mind to find reliable serenity in the woodlot.

Of course it doesn't always go well. Sometimes a tree chosen for cutting because it looks peckish, no longer a sturdy member of its company, falls with a distinctively hollow thump and shatters, revealing a punky interior so far gone in corruption as to be worth no more work. (A fireplace fire is contemptuous of punky wood, burning it sullenly, as a reluctant duty.) Occasionally the felling goes awry, from miscalculation of lean or veering of the wind during the wavery moment when a tree has concluded to fall but has not yet fixed on direction. It may maliciously slant off-course and lodge, half fallen, propped by the upper branches of a neighboring tree. This is a woodcutter's embarrassment, impossible to explain to anyone who happens by as something intended, and impossible to abandon, being a deadfall peril. If it cannot be jounced free, the choice is either to fell the second tree too, or to dislodge the hang-up by rolling it or by using a log chain to drag the butt out until the top breaks free. The work is arduous, a little dangerous, and thoroughly frustrating. On these infrequent days when malign spirits flit through the woods to perch on branches near by, all woodcutting can grow cranky. With each bucking cut the trunk rolls to new positions of inconvenience, and the saw binds inexplicably, and the axe glances wickedly, and the footing is precarious while you are trudging with a heavy four-length log on your shoulder. It is prudent to gather your tools and depart on such a day. Don't even stop to split what's been cut: the first piece will simply swallow three wedges, and smile.

¶ I CAN REMEMBER several times when, for more than an occasional day of spitefulness, woodcutting lost some measure of its attractiveness for me. Once was after the New England hurricane of 1938, before the Weather Bureau had begun its tactless practice of using female names as identifiers of natural disasters. Spinning viciously from the south, unheralded by orbital scouts then uninvented, this storm uprooted hundreds of thousands of trees in New England, and scores of them at the place where my family lived. With uncharacteristic unanimity the family nominated me for the task of cleaning up, since I was unemployed and given, it was observed tartly, to lounging around. The blowdowns were big trees, mainly oaks and beeches, but with a number of those giant straight-trunked pines that would have gladdened the eye of an 18th Century shipwright. The trees were not simply flattened but jackstrawed into tangled mounds, blocking the drive and entrances with mazy jumbles of leaves. One large oak propped the house, trickily wedged between dormers, and on the lawn two big oaks were uprooted, their dirt-encrusted underworks raised like giant fists.

Tidying up the place took months, limbing and sawing, tending great brushfires in the fall rains and splitting the yard-long logs that fitted the fireplaces of that house. I learned to work steadily, unhurried but unresting, and calluses replaced blisters as the capacious wood-storage area in the basement was filled rafter-high and the overflow supply of cut, split wood extended like a Roman wall outside. None of the fireplace wood was pine or other softwood, for we were believers in the old faith that conifers were not fit for a civilized hearth. There were cords and cords of oak and beech, and much of it was still left years later when, from the erosions of death and marriage, the family grew diffused and the place was sold. When the hurricane damage was finally cleaned up, I had some private satisfaction at a new competence in woodcutting, but if anyone had told me that getting in firewood had spiritual rewards, I would have been tempted to split him on the spot, deftly.

The considerable work of cutting and splitting was nothing compared to the problem, part physical and part emotional, of getting rid of the two uprooted oak stumps on the lawn. They soaked up labor like a sponge, and they returned an implacable hostility to my efforts. It would have been sensible to have had them bulldozed out and trucked off to a dump, but that didn't occur to me until so much effort had been invested as to make surrender unthinkable. Unlike the big pines that had, in thundering down, hoisted up large discs of shallow manageable roots, those stubborn oaks exposed giant onion-shaped masses of hardwood, braided over soil, gravel, and rocks, matted like the bows of tugboats. Each stump weighed tons and was utterly unassailable. Once the trunks had been sawed off at what had been the grade-line, there was simply no way to attack them with tools. They were too vast and convoluted for sledges and wedges and wholly impervious to axe and saw. For fruitless hours I gnawed at them with pick and crowbar, trying to remove enough stones and dirt so that parts could be sawed free. This proved hopeless, a work of years, for even when a single root was picked free at one spot it would promptly curve back and dive into the tangle again.

Then Mr. Kirkudbright came by, an elderly sage who had retired to a final career of closely watching others work. He prescribed boring holes in each stump, filling them with crankcase drainings, after which, he predicted with rheumy conviction, "they'll burn to flinders." I had not then learned that the predictions of neighborhood wisemen are not, in general, infallible, and I spent more hours, as well as several tool-bits, in boring holes in one of the monsters, which proved to be only slightly more penetrable than a bank vault. Crankcase oil did not disappear absorbently into the holes but simply sat there, taking on the stubborn character of the stump. I built a hot fire around the stump, kept it burning for hours, and found afterward that the stump was totally unchanged, except that it was now sooty as well as gritty.

In frustration and anger I left the monsters alone for a few days, going back to bucking and splitting the blowdowns, something where effort correlated with results. I remember glaring at the stumps in the misty afternoons of late fall; they had turned into beings, giant pagan idols named Gnarth and Hringar. Suddenly it seemed desirable to move the stumps together, so that a single fire between could eat slowly at both. It proved an immense undertaking, mining beneath to hack free remaining roots amid showers of dirt, charcoal, and crankcase oil; and then engrossing complications with chains, timbers, crowbars, and heavings on a block-and-tackle to inch Gnarth over next to Hringar. Finally they were side by side, and I built an artful fire in the irregular slot between them, using pine and spruce for flame, oak and hickory for incandescent heat, and liberal bastings of crankcase drainings for luck. During a stretch of sleety weather in early December I kept the fire burning for days, stoked with logs at night and built up to a blaze each morning.

For almost a week nothing happened to Gnarth and Hringar, and I began to wonder if the evil idols had contrived to interpret my fires as altar rites of praise and submission. Then one morning I noticed that Gnarth somehow seemed a little less huge. Both were still black, ugly, and defiant, with no identifiable areas of burning, but it was possible that there was some slight shrinking. After another week of steady fires their melting grew unmistakable, and the monsters could be edged closer together with just a crowbar. Not more than three weeks later the once-implacable Gnarth had disappeared forever, and Hringar had diminished to a charred and shapeless object that could be carried indoors in my arms and consumed in the fireplace on Twelfth Night.

The other time when the pleasures of woodcutting were somewhat abated for me occurred about a decade later. I was living then with my Southern-born wife and small child in a rented dwelling, converted from a barn, on the fringes of exurbia. We'd signed its lease on a golden September day when the place had appeared a treasure, with picture windows overlooking field and woods, paneled interiors, a great fieldstone fireplace, glittering new fitments in kitchen and bath, an unconventional but obviously advanced heating system. All details about the place were pleasing, an old fashioned flower garden, Dutch doors, a bathtub whimsically sunk until its rim was almost level with the floor, a round post of a tree trunk that rose surreally from the middle of the living-room floor to support the adze-hewn rafters overhead. It wasn't until we moved in that the topological constraint imposed by that whimsical post grew clear: the only way to put down the living-room rug was to cut a ten-inch hole in the middle of it, as well as an access slit in from one edge.

The first indication that charm had its costs came one October afternoon when it was observed that parts of the Sunday paper dropped somnolently from the sofa did not, in fact, reach the floor but instead arranged themselves flat against a nearby wall. The barn wasn't just drafty but utterly pervious, and hospitably invited in every breeze, from soft airs to fall gales. My wife, whose upbringing had encouraged the opinion that any temperature below 72 degrees Fahrenheit was a risk to health, sought out the landlord. (We learned much later that he was an entrepreneur who had arrived long before from some Mediterranean island, and that no great success had marked his career as a contractor until he discovered a special talent for tricksying-up rural properties for the enticement of innocents. While no one in the village particularly liked him, he was appreciated for his entertainment value, especially for the zest with which he litigated with tenants.) My wife was able to win no more from him than a brilliant smile. "Not mind if you want to chinka her up some," he said.

We chinked industriously all fall, an hour or two every evening and great spans of time on weekends. Twice we invited in friends for collective barn-chinkings. Also I used the expedient, common in northern countries, of setting out a band of hay along the line where the building met the ground, retaining this thermal bandage with staked boards and tarpaper. As the year wore into winter it became evident that the old barn really wasn't chinkable. Its ancient board-and-batten siding was so wonderfully weathered, worn, and warped that it wasn't so much a structure as a charming aggregation of apertures. We found that when the north wind blew, the best that the advanced-looking heating system could do, pulsing unceasingly, was to keep itself from freezing, with virtually no therms left over for us. The sunken bath, being embedded in frozen ground, could be used only during the fifty or fifty-five seconds that it took to chill water from overhot to overcold. Around Christmas we sought out the landlord again, finding him preparing his annual migration to Florida. "Just send checks to bank," he said. "If chilly, all right to bring in some from woods.

Only dead trees, you know? I been meaning to get woods cleaned up anyhow."

So began a winter of serious woodcutting. In the brittle January time when the days grew longer and the cold grew stronger, I spent every Saturday and Sunday bringing out enough firewood to last until the following weekend. It was a race with the lead constantly changing, the woodpile building a little when the cold let up, and dwindling down to a log or two when the wind scoured down from the northwest and the stars shone close and very bright in the winter heavens. The woodpile became a recording thermometer, graphing the differential effects of weather. An ice storm or heavy snowfall, for all that they complicated life and made it less pleasant, did not demand firewood nearly as hungrily as the cruel windy clear weather that followed a storm, when needle-sharp knives of cold assailed anyone straying from the radius of comfort around the hearth.

The woodlot began on a rocky slope a hundred yards from the house, a sparse stand of mixed hardwoods. Some forgotten misfortune—disease or a sequence of dry years—had killed some of the trees, and they stood unbranched, their bare trunks grey against the sky. These I chiefly cut, not so much in response to the landlord's admonition as by preference. A dead tree that is still standing and has not yet gone to punkiness is partly seasoned, burning without the heat-niggardliness of green wood. They were medium-sized, up to two feet in diameter, which minimized the need for splitting. Big firelogs can't be carried in armloads, and they have to be eased down to keep from smashing the andirons, but they can make a glorious blaze, and one that needs minimal tending. Looking ahead to the chance of hamperingly deep snow, or some temporary disability such as a sprain or a cold, I left untouched a few particularly desirable trees at the edge of the woodlot nearest the barn.

This reserve of emergency firewood was a symbol of that winter, which turned out to be an exceptionally cold one. Looking back now, I think I clearly envisioned a possibility of hobbling out during some week of need and hacking at those reserves of fuel. This behavior would not have stood up to rational scrutiny. We were not then so hard up that we could not have bought a cord of firewood, or torn up the lease and moved to a more habitable dwelling. Nor were we ever greatly uncomfortable, so long as we stayed near the big fireplace—least of all I, traveling five days a week by steam-heated train to steam-heated office. But rationality presents itself on many levels, and it had somehow become extraordinarily important to make it through to spring without giving up on that damnably charming barn. As for buying firewood, it would have seemed sinful to buy it when there were trees near by to cut; no one who was reasonably aware during the Depression years can be sure he escaped invisible scars that may, long after, prevent him from spending a few dollars that he doesn't absolutely have to. In a way that winter was like a game that had begun in good spirits and then gone deadly earnest. I remember thinking when working in the woods at weekend dusk that this was what it must have been like during the centuries when a steady fire was critical to a settler's survival. On other twilights I imagined that this is what it will be like when the great walls of ice once more creep down and man will have to struggle to dwell in small outposts above 40 degrees North anywhere on the congealing planet.

Much can be learned about the management of a fireplace fire when you are dependent on it for more than *Gemütlichkeit*. You discover that in a cold room a smolder of a fire has almost no value. Even a small fire, a couple of log-ends flickering in a modest blaze, is of use only when you can allow the circle of chill to close in, as when reading late at night in a wing chair drawn near, feet propped up. You learn once again that a spark screen is a heat-robbing abomination. The upper as well as lower limits of fire size become evident. A four- or five-log roarer will extend the arc of comfort somewhat, but it will also extend the too-hot-for-comfort arc and gobble logs in immoderate measure. A three-log fire is an excellent compromise, giving comforting warmth in a decently large habitable area without depleting the woodpile at a profligate rate. Your sense of timing sharpens, instructing you that the correct time to add another log is a short and variable interval before its need becomes apparent. The sensitive adjustment of the chimney-damper grows important, opened up a bit when the fire is being accelerated, and then closed down to the narrowest setting that can be maintained with regard for the speed and direction of the wind outside, without inviting little tears of smoke to weep in from the upper corners of the lintel. (The fire that smokes just a little is not necessarily bad: a fragrance of woodsmoke is associatively warming.) You learn to let ashes in the hearth accumulate to a broad, deep bed, leveled off almost as high as the bearing-bars of the andirons.

A kind of rough simplification sets in when a fireplace is used for heat rather than ornamentation. Accessory baubles like bellows and lighters, brushes and shovels, fancy matches and canisters of colormaking chemicals are put aside as trivial. Wood is chosen, cut, and carried in solely for its heat-giving property, which means heavy hardwood logs cut as long as the fireplace will hold. The andirons are spaced well apart, so that wood burned through at the middle falls inward, rather than charring into conical heels that tumble outward and smolder uselessly. In the winter of the barn, when the supply of wood outside grew alarmingly low, I'd sometimes carry in double- or triple-length pieces, too big to fit the fireplace but still perfectly usable when slanted in from each side of the hearth against a husky backlog. A merit of this rude arrangement—often used by European peasantry, and akin to the "star fire" (three logs, 120 degrees apart) of American Indians—is that it is in some measure self-throttling. A fire burns briskly so long as wood is kept nuzzled together at the apex, and it dwindles back to an economical idling condition as the wood burns away. This means that if you leave for half an hour the fire slows down automatically, and yet on your return it is necessary only to kick each log inward for a blaze to spring up once more. Of course it gives the room a disheveled look to have logs protruding from the fireplace like tongues from panting dogs, but it saves two or three bucking cuts per log, and the self-regulating aspect is an effective way of extending the combustion of wood hard earned.

By measures like these—expedients that conduce to a kind of flannel-shirted, stiff-shouldered, rough-handed, booted way of life—we came through the winter of the barn. It was in March, in a three-day storm that began in wet snow and went to rain, converting the woodlot to a foggy, wet ice-scape, that the woodpile drew decisively ahead of the cold, never to be headed again.

¶ IT IS SADLY TRUE that for many fireplace owners, living in treeless suburbia or cities, firewood must be bought rather than cut. Sometimes it is obtained from Joe, doing business from a classified ad that also offers yardwork, septic tanks emptied, exterior painting, satisfaction guaranteed. Joe can be all right, with accommodation developing over the years, and for all his imperfections far better than doing without. Persons of little fireplace sophistication, likely to wish a fire as animated decoration on festive days, somewhat like a potted poinsettia, can fill their needs from a variety of sources such as roadside nurseries and garden shops that do a brisk trade in cast-concrete statuary, blue crystal balls, and painted gnomes. Firewood is often for sale at highway vegetable stands, adjoining the piles of pumpkins, jugs of chemically inerted cider, and swags of corn ears and grasses that provide a more decorous fertility symbol than a yard dotted with bent tricycles. It is even possible to buy firewood in grocery stores and superdrugstores, the kind of spacious fluorescent marts that also offer aluminum chairs, color-TV antennas, and bagged charcoal in addition to more conventional apothecary sundries. Here firewood comes in the form of eerily smooth and round cylinders evidently ejected from an hydraulic press and packaged in containers bearing printed encouragements about freedom from mess. I tell myself that there is no call to be mired in bogs of intolerance about these machinemade logs, so queerly uniform, so far from anything that ever bent in the wind and sprouted green leaves, because giving in to disapproval also logically requires similar frenzies about hamburgers, frankfurters, and other assemblies of ambiguous particulate material.

In a number of cities firewood can be bought at the door, vended by rustic entrepreneurs exploring the boundaries of the capitalistic system. Responding to the ring, you will be offered some seasoned firewood, real good, burns great, just the right length, we stack at the side of the garage, just thirty-five dollars a quarter-cord. Frequently a certain amount of ducking and bobbing signals that, since you are so clearly a worthy person, some modest adjustment on price or quantity can perhaps be negotiated. At this point you are expected to go out to the raffish old pickup truck by the curb and verify the merit of the wood. Often it will be liberally muddied, a sign that a bulldozer got loose somewhere to advance the cause of progress by pushing over trees. But a moderate amount of mud doesn't affect combustibility, and you can even feel a flicker of gratitude that this wood, at least, has escaped the common contracting entertainment of bulldozing all trees in sight into a giant windrow, adding a few wornout tires as incendiary material, and then setting fire to the lot. If you do decide to bargain with such gypsy enterprisers it is well not to press particularly hard, because it is entirely possible to achieve a great triumph on the monetary side of the equation and receive an equal defeat on the amount of wood received. The four-by-four-by-eight-foot dimensions of a conventional cord can turn out to be an elastic cord indeed.

¶ A MAN fortunate enough to have access to a woodlot is doubly blessed in his chance to grow familiar with some fascinating and thoroughly rewarding tools. These are basically the axe and saw, but, depending on his toolishness, can easily extend to the peavey,* come-along, log chain, go-devil, glut, log dog, and that mad demon the chainsaw. In time this leads to equipment for the sharpening of tempered steel, including a wet grindstone, bastard and rattail files, a saw vise, a toothsett, whetstones, and handle-wedges, to say nothing of those boxes of oddments, not specifically needed but far too rich in promise of someday use, that collect in any

* A shrewd iron ingenuity first blacksmithed in 1858 by Joe Peavey, a gifted man of Stillwater, Maine.

serviceable workshop. The temptation to gather more and more of this delightful gear is strong, and it is well to remind oneself that the goal is to be a gatherer of firewood, not a lumberjack. In decades now at the edge of old men's memories, in the forests of Maine, Michigan, and the West thousands of men made their living with an axe, harvesting the forests as if it were ore and not trees that they brought out; living in bunkhouses and dragging loads of trunks in winter to lakeshores, riverbanks, and railheads, and leading the harsh, vigorous, skill-critical kind of life that can now be perceived only dimly. My context here is contemporary and wholly non-epic: the concern is of wood gathered modestly indeed, to feed the fireplaces of a centrally heated home.

An axe is a singularly honest tool, of such age and grace that you can be startled at finding it for sale in the prosy present. Occasionally axes may be glimpsed in the shopping-center hardware store, looking pale-hafted and unnaturally clean, back behind the riding mowers and mulcher-baggers, near the steak-charrers and patio torches. Their colorlessness makes them conspicuous among the brightly painted novelties around them, which are gaudied-up for precisely the same reasons that trade goods always have been. Axes have a quiet presence even in Sears Roebuck; they stand aloof and dignified, as if conscious of their immensely ancient lineage, descendants of the first tools man ever devised. Their peculiar asymmetric handles—except with double-bitted axes, which insist on symmetry—are not a selling novelty invented by stylists but are instead a fine example of the rational adaptation of a tool to human articulation. Just the presence of an axehandle is a continuing monument to the paleolithic inventor who, perceiving that a handaxe did its work by kinetic energy, then went on to the brilliant idea that a handaxe on a stick would have much more of this good thing that did the work. Even if he didn't generalize the principle, it was nevertheless a superb idea.

The beautiful steel heads on modern axes are deceptively simple, for a surprising amount of artifice is concealed in their forms. The curving thickness of metal behind the edge appears inconsequent, just a place where one shape is faired into another, but if you grind off much steel there, you will ruefully find that the axe has undergone a marked personality change, becoming a cranky, rhythm-breaking tool that jams in the cut and has to be tugged free. The hardening and heat-treating of an axehead is finicky, balanced between not-quite-hard-enough, which won't hold an edge in heavy service on hardwood, and just-a-touch-too-hard, which risks breakage and can't be filed. A good axe costs less than ten dollars even today and is a great bargain as a general-purpose tool, fine for clearing land, building shelter, procuring fuel, and serving as a kind of father-tool in creating a hundred necessities and conveniences for woods life. It wants to be kept sharp; a dull axe is just a club, a blunt instrument, and dangerous into the bargain.

More than most tools, an axe has strong aspects as a teaching machine. It becomes highly individual in use, instructing you in its reach and balance, the bite of its edge, what it can and cannot be expected to do in a single stroke, the corrections of eye and arm needed to bring it, with accelerating velocity and precise angle, to your exact aiming point. Most women, for reasons that must derive from childhood, are singularly inept with an axe, wielding it in gingerly fashion, with a swing like that of an unlubricated robot.* Not that an axe calls for Homeric flailing; it is a crafty, subtle tool, one that wants to be used with economical precision. One of the most skillful axemen I have known was a Maine farmer who, when I knew him, was a rather flimsy old man. His name was Walter, pronounced Wall-ter, and he swung his axe in a fashion that reflected seventy years of practice. He preferred talk to work, possibly because his only steady companions were some cows and pigs of not much account,

* Marguerite de La Roque was an exception. She was a young Frenchwoman of the upper class, wholly without survival-related skills, who was marooned in 1542 on a small island on the northern reaches of the Gulf of St. Lawrence, not far from Labrador. Cousin or niece of an exceptionally God-fearing Calvinist who was leading a colonizing expedition, Marguerite had a shipboard dalliance that, when discovered, appeared to invite divine wrath; she and her elderly nursemaid were put ashore and abandoned. As the ships hoisted sail to depart, Marguerite's lover dived overboard and swam ashore to join the two forlorn women. They had a few provisions, several arquebuses, and an axe. The lover died in the first winter, as did the baby born nine months later. The nursemaid died during the second winter. Marguerite killed three bears, at least one of which was a polar bear. After she had been on the island for twenty-nine months she was rescued—ragged, half starved, not quite mad—and brought home to France. She became a schoolmistress, and very pious.

and as he spoke he unfolded a powerfully barnyard bias on human affairs. When shamed by a continuing lack of conversational response, Wall-ter would pick up his rather light axe and cut rings around my friends and me. Being husky youngsters, we tended to swing in the style of Achilles having a tantrum outside the walls of Troy, but Wall-ter's swing, though it looked modest, contrived to use the axehandle as an accelerator that brought axehead to wood with useful velocity. It was never allowed to meet the grain in a square, 90-degree confrontation, but somehow always managed a slicing or diagonal downgrain cut.

Unlike us, Wall-ter was a marksman with his axe, each stroke landing exactly where it was wanted, not an inch away. (Precision is important with an axe because strokes that are off the mark fail to free a chip and are essentially wasted.) Typically we failed to start the sides of our cut far enough apart—they want to be just a little less than the diameter of the wood being cut—or we failed to establish the ideal angle to the sides, about 50 degrees, that is a nice balance between penetration and cutting ease. Finally, we were inaccurate, wasting strokes that landed an inch off the mark. The result was that Wall-ter, having produced chips the size of saucers and cuts as clean as if chiseled, would finish long before we did and launch forth into another anecdote about his scandalous barnyard. After this happened a few times we learned to treat Wall-ter, for all his manure-seasoned garrulity, with the respect that such artistry demanded.

Although a wonderfully versatile tool, an axe is inefficient on big hardwood trees, where a considerable volume of wood must be converted to chips at each cut. Here a saw, concentrating its work in a narrow kerf, is much more energy-effective. For generations of lumbering this meant the two-man saw, a limber steel band with teeth arrayed in a gentle arc that resonates very nicely with the swing of human arms. Its handles are vertical wooden posts, and one of the cardinal rules of two-man etiquette is that you take care not to bang your partner's knuckles into the trunk. Other commandments are: Pull, don't push; Don't bear down on your partner's work-stroke (because it will make him as mad as a hornet); and Don't pick up the tempo unless you are damn sure you have the endurance to bring your challenge off. Two men who work easily together can cut through a log in somewhat more than twice the time that one man can now do it with a chainsaw, which seemed wonderfully fast when we didn't know better. Now of course the reciprocal is evident: a chainsaw will outcut two men, be tireless, and afford absolutely no chance for small talk, barnyard or otherwise, since it blankets its vicinity with an unnerving mechanical caterwaul.

For persons unfond of chainsaws, several alternatives are available. One is the woodsmans's crosscut saw, using steel somewhat thicker than that of a two-man saw so it can be pushed as well as pulled. It is fitted with both a conventional handsaw grip and a detachable post so it can be used in two-man fashion when a collaborator turns up. Then there is the large family of framed saws, descendants of the wooden, turnbuckle-tightened buck-saw, that are now variously called Swedish, Finnish, pulpwood, bow, and D-frame saws. (The folklore in some localities attributes the Scandinavian names to an alleged preference of such immigrant stock for working alone in the woods, in contrast to the gregariousness of French Canadiens). These saws use the springiness of a tubular steel frame to secure both ends of a thin steel blade. Like the axe, they are as a rule so inexpensive as to suggest that entrepreneurs somewhere are nodding. Saws like these—light, strong, and sharp as the winter wind—are pure pleasure to use for limbing, and can slice through a heavy hardwood branch in a few incisive strokes. They are specially adapted for use with a saw-buck, with one foot comfortably propped up heron-fashion so the forward stroke has a downward thrust.*

If you develop a taste for cutting smaller thicknesses of firewood on a sawbuck (it is hard not to), it isn't necessary to lug the cumbersome thing off with you. Just drop a few spikes in your pocket when you go to the woods. Roll a short, heavy length of trunk to a level

* Does anyone know why having one foot propped is so satisfying a position? Once you start watching you can find it everywhere, on sawbuck, fireplace fender, bench, nail keg, desk drawer, stoop, bar rail, stone wall, and pickup-truck bumper or runningboard. Is it a territorial signal (this here is mah ol' truck)? Is it because it gives you a place to put an elbow or forearm? Is it a signal of nonhostility, because you can't draw a sword or swing a club with any degree of success while standing on one foot with the other propped higher? (Ah'm friendly; how 'bout you?) Is it, like stretching, a signal of ease?

spot, and cut four arm-thick branches about seven feet long. Sharpen an end of each and tap them into the ground so that they bridge the trunk in a pair of X's. It will prevent binding during sawing, and make you feel clever, if you space the X's at a distance that is just short of your fireplace length. Then spike each arm to the trunk. This gives you an on-the-spot sawbuck that can be made in minutes and taken apart in seconds. It is a trick showed me by my father, who saw it in use in the Canadian woods almost a century ago.

As to chainsaws, it is really not fair to dismiss them out of hand, like snappish, snarling dogs. At first use, a chainsaw seems singularly untranquil. Heard close up, the tongue that it speaks in is pure hysteria, and the ferocity with which a sharp chain eats through hardwood is chilling, the sawdust pouring from the kerf in a torrent. With horizontal cuts, as when cutting a stump off close to the ground, the din is unendurable when the exhaust comes nearer your ear, and the ringing afterward is an unpleasant reminder of the opinion of otologists that permanent hearing damage is possible. The racket adds some slight risk to felling, since it masks the cracking signal and thus deprives you of a second or two of lead time in scrambling away to safety. A chainsaw can also be skittish when the end or top of the blade is loaded, kicking up in your hands as if it were trying to get at you. Finally, a good-sized chainsaw can feel manageable enough at the beginning of work but, what with the noise, vibration, and tension, it gains weight astonishingly in half an hour's work, inducing in your arms a distasteful muscular flitter. It is a joy to put the damnable thing down and flip the kill-switch.

As you grow accustomed to a chainsaw, most of these drawbacks disappear. It is a handle-able tool of uncommon energy, eager to trade hard work for a cupful of fuel. It *likes* to work, the chain biting into hardwood with zest. It speeds through what would otherwise be drudgery, dropping off lengths of solid firewood in just seconds a cut. The kickback tendency becomes high-spirited instead of threatening, and the masking of cracking sounds during felling can be thought of, not very logically, as simply adding more spice to that exhilarating interlude. And while the full-throttle cry is never endearing, the friendly, hurrying *pop-bang* idle of a well-tuned machine grows almost engaging, a kind of tail-wagging, romp-ready signal. To most machine-minded people a two-cycle engine is inherently interesting, a sophisticated descendant of loutish and moody predecessors. Up to the 1940's—by chance about the time that chainsaws themselves first appeared—two-cycle engines were tedious mechanisms, heavy, sullen, a little gross. They started only after elaborate propitiation and many entreaties; they preferred to stop. Then, thanks to a bit of fresh engineering attention and improved materials, their personality changed and they became eager, light, almost manic. I think it possible that we live in a time of transition just now for chainsaws, and that before long they will become mannerly.

Until then, do not harden your heart toward the little demons.

❡ ASK A MAN who sets a good fireplace which tree gives him the best firewood, and his answer will reflect not only the timber stands in his township but also some of the absolute certainty that once marked sectarian faith. In northern New England you may be told that there is no question that the prize is rock maple or golden birch. In Connecticut and New York the favorite may be beech, ash, or a fruitwood like apple or cherry. In the midatlantic states and much of Appalachia the special treasure is often hickory or oak. As with most proclaimed certainties, such dicta are a little uncomfortable. For one thing, most of these trees occur throughout the over-all area, which sets you to wondering how a tree can be a prince of the forest in one place and nothing much a few degrees of latitude away. Perhaps variations in esteem correspond to subtle differences in local varieties. Perhaps the differences have been transmitted over generations of woodcutters from the times when the value of different wood was shrewdly assessed by native sages. It is also possible that in firewood, as elsewhere, belief modifies perception; and the degree of satisfaction derived from a beech or applewood blaze varies according to the cherished opinions of those who hold their hands out to it.

This is the sort of problem that ought to be speedily solvable in a laboratory. It ought to be, but it probably wouldn't. At some good-sized research institutions investigators would briskly set about establishing a Standard Beech, with full consideration of such factors as age, density, micro- and macrolocation, previous history, and group adjustment of the beech (from which random numbers of core samples were taken in scrupulous adherence to the protocols of the international understanding on coring beeches [Geneva 1967, with Albania abstaining]). This is preliminary to measuring calories per cubic centimeter when each sample was combusted for a Standard Minute in the sophisticated new computer-aided reactor named, as a matter of *noblesse oblige*, the Congressman Marvin Armbruster Standard Fireplace Facility, and identified during the ribbon-cutting as a key element in the vital competition for the minds and hearts of mankind. "Hopefully," the fourth quarterly progress report will say, returning to proximate English after some machine-readable alphanumeric banter with bibliographic retrievers, "if certain programming anomalies now being encountered are better understood in opportune season, the degree of confidence of acceptability of Standard Beech test cores will achieve Milestone III, and preliminary work can then be initiated on the very challenging task of provisionally defining the major parameters of the Standard Apple."

Some numbers for heat values already exist. A U.S. Forest Service table (overleaf) rates thirty common woods.

Species	Weight per cord, in lbs.	Gross heat value, per cord, in BTU's
I. *Extremely heavy woods:* apple, black birch, hickory, hornbeam, locust, white oak, blue beech	3890	26,800,000
II. *Heavy woods:* white ash, beech, golden birch, sugar maple, red oak	3400	23,400,000
III. *Moderate-weight woods:* black ash, white and grey birch, Norway and pitch pine, black cherry, elm, soft maple, tamarack	2880	19,900,000
IV. *Lightweight woods:* aspen, basswood, butternut, hemlock, willow, white pine, white cedar, balsam fir, spruce	2100	14,500,000

These values are for seasoned wood—defined as air-dried, under cover, for six months or longer. Freshly cut living wood is an inferior fuel because the considerable amount of water it contains has to be converted to steam, and the energy required to do this is subtracted from the useful output. Seasoning can make a difference of more than 20 percent in wood's heat value.

There is more to firewood than its specific heat value, which is a fortunate consideration, seeing that there isn't much you can do immediately about what stands in the woodlot. In my experience most of the hardwoods can produce delightful fires. It is in woodcutting, and especially splitting, that the differences are most evident. Hickory can be cranky to split, and maple only a little less so. Golden birch is a little easier, and it has a rewarding interior texture that almost obliges you to take off your work gloves just to touch it. Ash and beech are amiable on the splitting-block, and quite large pieces can be split with axe alone, not insisting on wedges. (Incidentally, green wood is usually easier to split than dry; and if you must split old wood, do so when it's frozen, and you'll have less of a tussle.) Most oak is grumpy and mulish during splitting, with a lacerating texture: it reminds you of the ancient proverb about how the best wood warms you twice, once when you split it and once when you burn it. A peculiarity of oak is its reluctance to season out of doors. If you try to counter this by bringing it inside to a heated shed or basement immediately after splitting, it reacts by giving off a pungent greenish odor, rather like a busy Paris *pissoir*, which promptly diffuses through the house. The smell lasts for weeks, and you can get used to it, although it does seem to make visitors noticeably uneasy.

The softwoods make the least desirable fireplace fuel, which is a pity because these are lovely woods, light, strong, a pleasure to touch and smell. They can be cut with relative ease, a properly set saw slicing through them in no time, and on a bonfire or in a woodstove they kindle easily and burn in a flashy, excitable way. But in a fireplace these woods reveal three disabling drawbacks: they sputter and spit like packets of firecrackers, they lack durable warmth, giving only a quick pulse of heat before fading, and in regular use they soot up a chimney badly. Their spitting pyromania, lobbing incendiary fragments about the room, requires almost constant use of the abominable firescreen, something that discriminating firetenders avoid except when they leave the room, or when a blaze has grown so vigorous that its infrared radiation needs for a few minutes to be throttled back by metal mesh. (A screen weakens and diminishes a fire, robbing it of presence, and those who use one regularly and unthinkingly deserve the pale, heatless fire they get.) As for their sooting the chimney, the worry is not aesthetic: it is that a seriously soot-clogged stack has a predisposition to catch fire. If you have never had a chimney fire, number the fact among your blessings. It is a roaring, spark-spouting scariness that comes on cold winter nights, the soot within the chimney itself taking fire. There isn't a great deal you can do about it—aside from keeping an eye on woodwork adjacent to the chimney at the ceiling and roof rafters, and doing what is needful to quench the fountain of sparks that shoots from the chimney—and often it burns itself out without causing damage. But its possibility, properly enough, discourages the heavy use of softwoods in a soot-prone chimney.

All of which is not to say that the occasional use of a little pine or cedar is utterly without redeeming social utility. Both are excellent kindling, useful in an outdoor fireplace and handy as an occasional spur to a lazy fire.

White cedar is one of the friendliest of the softwoods, light, strong, and fragrant. It displays its straight-arrow personality by behaving most agreeably on a fire, burning with a hot but not very durable flame, and not sparking quite as much as its coniferous relatives. Pine knots also have a place in a well-stocked woodshed. Tossed one at a time on a bed of coals some stormy winter evening, they burn with a bright distinctive light that is a playback of summer Suns. (Technically, it's because the knot is densely packed with channels of resin.) A pine knot can also be used to light outdoor chores some night when every flashlight on the place has sunk into orangey exhaustion. Of course any neighbor who chances to glimpse you brandishing a flaming pine-knot torch in the woods is certain to remember it, adding it to his estimate of your character that may, who knows, ultimately lodge smoldering in the F.B.I. files.

In my opinion, the grey birch is rarely worth the effort of cutting. It is close to being a fraudulent tree, good-looking enough to deceive the innocent but in fact giving neither heat* nor durability. It belongs only in fireplaces that are stacked for show but never lit, or as ornamentation for one of those poignant ersatz blazes constructed around moving colored lights.

Greenness in at least part of your woodshed holdings is not all bad. A big unseasoned backlog can be used to advantage at times when a bright but not particularly

* Walter Needham, a Vermonter with a keen sense of country ways, tells of a man burning green grey birch in a stove: "He had to look inside once in a while to see if the blaze had froze."

hot fire is called for, which are most days in the prevalence of central heating. Also a green log may be used as a moderator on a fire that has chanced to grow too energetic, serving this purpose better than the abominable firescreen, if you don't mind the hissing noise as sap is driven bubbling from its ends. So for these purposes, as well as for extending a dwindling stock of dry wood, having some fresh-cut sticks available is convenient. For most men the woodshed plan is to burn last year's cut, along with any leftovers from the year before, and to set aside current replenishment for next year's fires—an approach that endows the woodshed with a reassuring feeling of continuity, as though it were orbiting the Sun on its own individual ellipse. But it is not in the natural condition for sheds to remain tidy, particularly if there are several people adding and taking, not all of them clear about, or even aware of, the master plan, and at least one of them given to picking over the separate piles with something of the same frenzy that might be devoted to a table of blouses on special sale at Tudbury's.

Occasionally when insomnia drops in I spend time in the dark arranging details about the woodshed that I will have in the back of that someday house on the hilltop in Maine. It will be a long, floored structure like one kind of stable, with sliding doors that won't swing against deep snow, and a gentle ramp, with no step, for easy admission of a wheelbarrow or woodcart. It will be accessible from within the house: no walking in the snow in slippers. I haven't yet decided if the shed will be heated—a convenience but not essential—but there will be a switchable exhaust-fan for the oak.

In this building I will keep the wood that I will cut and split for an hour or two every day, maple and golden birch, oak and hickory, beech and ash for weekdays, apple and black cherry for occasions. The wood will be sized for different fireplaces—husky 40-inch logs for the sociable fireplaces and 28-inchers for the smaller fireplaces in the study, baths, and bedrooms. (In the later months of the year, I shall frequently go to sleep with firelight on the ceiling, to match the aurora borealis flickering in the Northern sky.) The shed will be fitted with partitions, like horse stalls though of different proportions, and in this way I'll keep firewood separated by length of cut, kind of tree, and vintage. It will be as discriminating as a wine cellar, stocked with resources from which suitable fires can be drawn, variations on the basic fire in Someday House, with an oaken backlog, maple for substance, applewood for fragrance, with a bit of driftwood from the coast for blue- and green-flamed garnish.

TENDING FIRE

GETTING THROUGH life seems to require an uncounted number of small skills. It begins as we learn to feed and dress ourselves—those shoelace and necktie knots!—and ends with the delivery on our deathbeds of an adequately Delphic remark. In between, our days are spent in efforts to perform small co-ordinated acts with competence if not grace: cooking, parking a car, earning a living, making music, balancing a checkbook, conferring civil inattention upon strangers, repairing damaged objects and feelings, responding to the friendliness of children and dogs, painting window frames, eating boiled lobster, ignoring tactlessness, tightening loose screws. There is no end to it.

Sometimes the deft performance of modest skills can grow to the proportions of a private vice, as with a magazine photographer with whom I once traveled. Like many in his occupation he had a morose cast of mind, tending to dwell on the probable breakdown of his equipment, the difficulty of his assignment, the malice of the weather, the caprice of his art director. I was surprised to discover that this dour man had a secret compassion for, of all things, motel television receivers. "It's the most maltreated mechanism in America," he told me seriously, "the victim of every idiot knob-twister that comes along. You find the things with the contrast up full, linearity all shot to hell, rolling like a ball. I think people throw shoes at them." In one camera case he carried several of the noncapacitive fiber wands that TV repairmen use for adjustments. Every night in a new motel, even though we might have driven all day long, he'd get out his tools and, delicately touching up the back-of-the-set controls, adjust the receiver to the fullest tone scale and most exact linearity. "That's better," he'd say finally, putting away his wands. "Not good, but better." He loathed television programs, never watching them out, but all across America he left a trail of sensitively adjusted TV sets, like an electronic Johnny Appleseed.

Something of the same loving nicety can be seen in the way talented people manage a fireplace fire. Usually they do it so inconspicuously that the process is very nearly invisible. Such a man abhors Wagnerian scenes and would never flail at a fire with tongs or poker, never drop a large log atop the structure of a fire as gracelessly as stew is dolloped out in a chow line. You will almost never see him use bellows to stimulate a laggard blaze, partly because his fires will not have been permitted to be laggard, and partly because bellows are likely to floof up a nuisance of fly-ash. He chooses a new piece of wood from the fireside basket with an eye to its cross-sectional shape and he fits it into the fire like a fieldstone into a wall. Often he adds the new piece a few moments before its need is obvious, knowing that a fire fed only when hungry will have a jagged temperature chart. Watch his use of poker or tongs to achieve a settling of the logs until they lie comfortably together, re-inforcing each other to feed the flames between, and the way he returns to the blaze any burnt-through pieces that have dropped aside. These are all quiet and adroit attentions, very unlike the thrashings of an inexperienced firetender who is all too likely to treat his blaze as a victim to be thrashed into acceptable behavior.

The complex relationship between a man* and his fire has something of the relationship of actor with director, pilot with plane, parent with child. The closeness of the linkage is reflected in the ancient taboo against touching another man's fire. The old taboo has transmuted itself into etiquette, and today as a guest in another man's house most of us would be as likely to kick his fire as to fondle his wife, at least in his presence. This means that if one's host has constructed a smoky miscreant blaze, awkwardly assembled and behaving like a petulant child, it must be accepted just as the child would be, as one of the penalties of accepting his hospitality. The most that can be done, when he is out of the room for a moment, is to give it—the fire—a remedial kick or two, just so long as it will not look greatly different on his return. One of the indicators of a genuinely close relationship comes when you are made free to tend another man's fire; it

* The gender of the tender is without significance, for women can be exceptionally artful fireplace managers. They also often display high competence with the draft and damper of a cookstove, showing skill much beyond what might be expected to result simply from daily practice.

comes late in a developing relationship, about the same time you feel free to rummage in his refrigerator.

The firetools in our decorator-dominated times are often elaborate and a little silly, traveling a course divergent from reality. Consider for example the ash shovel, now typically an enlarged version, in brass and black iron, of the toy provided to sandbox toddlers. In the 17th Century cooking fireplace a shovel was a working tool, needed to aggregate hot coals under a suspended pot or kettle. It also had a specific use if food was to be fried or grilled. Coals would be withdrawn from the main fire with the shovel and spilled out on a bare spot on the spacious hearth. An iron trivet was put on this subfire, and a skillet or spider put on the trivet; after the cooking chore was done, the shovel returned the coals to the main fire. The shovel had a third use, in removing the embers of the fire that preheated the wall-oven for baking. Nowadays the shovel has none of these functions, and its presence is a dangerous temptation to housekeepers not fully instructed in the folly of regarding ashes as messy and to be carried away on sight. Today the shovel is needed only for the infrequent task of leveling ashes to an even bed, and this is something that can be done with a sweeping movement of poker or tongs.

The poker is another tool in the contemporary "fire set" that has only modest usefulness. Not that fires don't need poking; they thrive on it, the way a house dog needs patting. But unless there is a light-colored rug near the fire, a great deal of deft fire-poking can be done with the toe of a shoe; and when that is not practical, tongs pinched together make a fine pusher. A poker fitted with a side barb has moments of usefulness, as when a big backlog needs to be rotated, although this also is well within the capability of tongs. Unless the fire being served is a great roarer of heavy logs, tongs serve as a general-purpose tool, the only one needed. For obscure reasons, possibly tradition lightly dusted with stupidity, designers often provide tongs with a single handle located outboard of the pivot. If this excrescence is innocently grasped, the tongs thereupon insist on two-handed operation, which is about as rational as two-handed use of a gearshift lever or toothbrush. Most pairs of tongs adapt very nicely to single-handed use, rather like a large pair of chopsticks that move deftly about the hearth making tweak adjustments in the fire. The first time you pick up a new pair of tongs it is prudent to see if the tool has been shaped in a way that can pinch fingers or palm; it is not yet a law that he who designs a defective tool is sentenced to use it.

During a walk in the woods it can be a pleasant pastime to keep an eye out for a natural pair of fire-tongs. It may appear that wood is not a material of choice for a firetool, and so perhaps it isn't; but wooden firetongs will survive for several decades of regular use, which is not bad longevity as everyday tools go. Your quest is to find a wishbone structure of limber branches, perhaps an inch or a little more in diameter, with legs that bow slightly toward each other. Often a fork of this shape will be found in a thick clump of brushwood, with each piece competing hard for the sunlight, or in the sucker growth arising from the stump of a tree felled several years ago. Another place of good hunting is along the trunk of a tree blown down by high wind or felled in an icestorm, a tree that has nevertheless lived for a few seasons with its trunk parallel to the ground, long enough for new growth to adapt valiantly to the strange new position of the sky. Look for a wishbone with legs six inches or so apart, just limber enough so that one hand can spring them together. A natural pair of tongs like this is often suited to a country-house fireplace, and a fitting memorial to its parent tree. After long use, the ends of the tongs will char into fire-hardened tapered points, not unlike those on ancient spears.

Although not firetools, andirons or firedogs and iron firebaskets or grates are fireplace adjuncts called for by the conventions of our times. You can do without them simply by building your fire on the hearth, nestled among the ashes of previous fires. To achieve reasonably even burning along the length of logs, it is convenient to build the fire an inch or two above the ashes, using two firewood lengths as andirons; green wood banked in ashes will last for days in this service. If more durable supports are wanted, a pair of low piers made of bricks or flat rocks will function well for years. Such extemporized andirons are quite satisfactory, serving all but two of the purposes met by traditional ones. They provide no buttress to keep a round upper log from rolling out onto the outer hearth as a burning fire re-arranges itself. (This can do little harm except when a fire is left unattended, and can be avoided simply by not capping blazes with round logs.) The other drawback is visual: a fireplace without andirons can look oddly wrong, like a familiar person without his everyday glasses. We can be so accustomed to having the proscenium of fire framed by gleaming brass or black iron that their absence may seem queer. It is like returning to a place known in childhood and perceiving not just smallness and familiarity but also some strange transparent holes in space caused, one realizes after puzzled thought, by the absence of trees or buildings now entirely gone.

There are people who burn their fires on cast-iron or welded grates or baskets that confine the wood at front and rear, and sometimes even at the sides. It must be admitted that grates and firebaskets make good sense with particulate or lumpish fuels—anthracite, soft coal, cannel coal, peat, or dried animal dung. In fact it would be inconvenient to burn these substances without a grate to keep the fire coherent and to allow access of air from beneath. The idea of an iron firebasket is very old of course, tracing back to the tripod braziers of many Mediterranean peoples, to the iron firepots set in sand on early sailing vessels, and to the great metal baskets flaring with resinous, smoky flame that Greeks used for illuminating public squares at night, for watchfires at mountain passes, and for beacons along rocky coasts and harbor entrances. One can hardly contend that anything with so grand a heritage has no place in a wood-burning fireplace. The fact remains, nevertheless, that bunching lengths of wood together in an iron cage has more convenience than charm, the convenience being achieved not so much by making skilled firetending unnecessary as by making it impossible. Reluctant green wood, for example, simply smolders mulishly in a firebasket, swallowing armloads of kindling to little effect. On firedogs the same wood can be assembled in a structure of triangular section where a little kindling can create an internal fire that dries the logs as a preliminary to a fine frisky blaze.

In England the iron fireplace grate became important during the 16th Century, when the nation—as fireplace-dependent a country as any on Earth—ran shockingly out of firewood. The circumstance may be instructive for our own times. Naturally firewood didn't disappear overnight, and remained available to those having access to private woodland: to farmers, the gentry, the landed nobility. What apparently happened was that the forest land open to entrepreneurial lumbering became less and less productive as it was repeatedly cut over, and wood had to be carted farther and farther to bring it to the cities that consumed it in enormous and growing amounts. It was also more profitable to use new growth for lumber rather than fuel, so that firewood, mainly from leftover cuttings and unsuitable trees, grew scarcer still. Growing competition for reduced stocks of wood fuel also came from the booming glassware industry. When King James I discovered that a single glassworks was burning 400,000 pieces of wood a year to get potash, by royal proclamation in 1615 he prohibited the practice.

As a consequence of all these pressures the cost of

firewood climbed so steep a curve as to convert what had been a necessity into an impossibility. In London the cost of firewood rose 800 percent between 1531 and 1632, whereas coal (used in a minor way for heating ever since the Roman occupation) remained relatively cheap. This was quite enough to dissolve prejudices against the "new fewl," which was resisted initially because it smelled strange and gave less light than a wood fire. With the appearance of smaller, shallower fireplaces, fitted with coal-adapted iron grates, and as the simplicity, concentrated heat, and staying power of the new fuel became recognized, the coal fire became the primary space-heating method for England. It was not in the nature of the 16th Century mind, as wide ranging and imaginative as it was, to make nice distinctions between renewable and nonrenewable resources, nor to anticipate what many hundreds of thousands of small soft-coal fires would ultimately do to the microclimate about English cities. Thanks to the supple adaptability of taste, is was not long before the coal fire began to develop favorable associations and then an aesthetic of its own. The mound of even, red-glowing coals surmounted by delicate blue flames grew much appreciated as comforting protection against raw, wet weather outside.

The attitude has continued for centuries. A Government official of my acquaintance tells of a marvelous coal fire that glows from October to April in a sedately dignified Baltimore club. "After I began going there occasionally for lunch, my eye was caught by the rich, quiet perfection of the fire in the paneled hall. It was the epitome of an English coal fire, always beautifully behaved, although I never saw a houseman tending it. One day a horrid thought came to me. Taking a moment when no houseman was present, I dropped to my knees and looked up at that fire from beneath. Sure enough, there was copper tubing snaking up to that beautiful fraudulent thing, with gas flames heating fake coals of carborundum. It was pure counterfeit, made by some sly Anglophile craftsman."

A similar example of ingenious deception has occurred as a by-product of space research. The plume of flame from a big rocket using high-energy fuel creates temperatures compared to which the heat of a roaring woodfire is merely a baby's breath; and this (along with even higher temperatures generated by atmospheric reentry at ultrahigh velocities) has required intensive research for materials capable of withstanding extreme heat. I have to report that several engineers at one rocket establishment have formed a private, entirely legal outside company to manufacture fake fireplace logs from one of the new refractory materials. First they took a few true logs, complete with bark, knots, and other marks of genuine wood that has been watered by rain and warmed by the Sun. From these models they made molds that accurately reproduced every surface detail, and in the molds they cast a mixture of the new materials, creating false firelogs of preternatural durability, for sale to persons adaptable enough to find solace in a gas-fired charade. It is better, I suppose, than those eerie electric flickerers. As to the makers, it is conceivable that God will forgive them.

¶ WATCH A MAN build a fire and you will sometimes see more of his nature than perhaps you should, or at least more than he might prefer. Some men bring a brisk competitive antagonism to the task, challenging the son-of-a-bitch to *dare* to smolder sluggishly. They strew on a week's supply of kindling, pile on enough wood to begin an ox-roast, and as a consequence are shortly obliged to open doors and windows to keep the room from becoming an oven. Others bring the healed scars of past difficulties to the task, setting out the wood meticulously, placing each piece of kindling just so, adjusting the chimney damper to an imaginary nuance of openness, tucking crumpled newsprint beneath in exact pattern, and checking all details carefully before the match is struck. These clearly are men who believe that the capricious spirits of the hearth are best wooed with an exact attention to ritual detail. They build not so much a fire as a self-fulfilling prophecy, never testing the possibility that their fires would start quite as well with a quarter of the preparation. Finally, there are originals, men with a stubborn independent streak who believe that conventional methodologies are wrong and that their own procedures, worked out with brave disregard of convention, are true contributions to human knowledge. These are the men who construct fanciful little teepees of kindling, or lean wood against the back wall of the hearth, or construct precarious log cabins of firewood enclosing nests of kindling. I knew a man who somehow concluded that other people always built fires upside down and that the only proper way to do it was to put logs on the bottom, heavy kindling in the middle, and twigs and paper at the top. He was a gentle academic, with only the foggiest awareness of the physical principles involved, but since he was blessed with a fine woodshed of seasoned hardwood, his eccentric inverted fires usually started. "See?" he would exclaim. "Didn't I tell you?"

A key to effective fire-starting, and to deft tending once the fire is under way, is concentration and control of heat. Except with uncommonly dry wood, and except with big fires that are hot enough to burn almost anything, a fireplace fire marches best and is most readily controlled at the marginal level where a diffusely assembled fire would drop back to a smolder and yet a well-organized one blazes merrily. In nuclear parlance, a docile fire is one that is just critical. A fire that is overkindled or that is fed too much dry fuel ceases to be a friendly domestic amenity, growing willful and fierce, controllable only by choking it with ashes, or by disassembly and starvation. These are measures a careful firetender dislikes employing, since they signal that for a time at least the fire has taken charge. One occasion in the past when an extremely hot fire was desirable was the drizzly fall day when it seemed wet enough to burn brush in a clearing in the woodlot. (This was in the receding time when an outside brushfire was not an offense against the laden air.) Often the project came up on a damp windless day at the end of an autumn storm, and everything would be so soggy that even a highly principled woodsman might think for a moment of using a cupful of kerosene to help get his fire started. But this tempting vulgarity would be put aside, and in fact it took only patience and a little art to nurse a flame into independent life. It built up slowly from its nest of birch or cedar bark, accepting selected tidbits of dead twigs as a convalescent is coaxed to take nourishment. It would grow moodily into a circular fire several feet in diameter, fed with wrist-thick windfalls from the forest floor, pieces that had shed bark to become grey-black wooden bones.

Even though it was still an inconsequent fire, you could see it begin to become greedy, snatching up rust-colored pine and hemlock boughs, beginning to eat its own smoke. Slowly it began to sound like a fire, first with individual crackles and hisses but then growing toward the distinctive summing *swoosh* of a bonfire. When the time arrived to drag up and toss on the first leafy branches remaining after firewood had been limbed and bucked—brush still so sap-filled that it would sprout shoots if left at the edge of a brook—the fire would seem to pause a moment, deceiving you into thinking that it had been given too much too soon. But then it would burn centrally through its leafy roof, bursting out with a single columnar flame of almost scary energy, carrying sparks and ash high in the air.

More and more brush would be added, cautiously now, until the fire stabilized as a white-floored circle nine or ten feet in diameter, so intensely hot that it would consume any wood that you dragged to it, so fiercely radiant that it beat against your face when you approached near enough to toss on new branches, so oppressive that only the damp autumn drizzle kept you from stripping off your shirt. Burning brush was not a thing to be undertaken on impulse, for it had the solidity and obligation of a day-long task, begun with the challenge of lighting a fire from damp materials, carried out with the attentiveness that a large fire insists on, and then tapering off in several reverie-filled hours of seeing the fire properly out. It was never comfortable to leave your brushfire until it had died back to a circle of soft ash darkening in the late-afternoon rain.

These pagan bonfires of brush are entirely out of fashion today, of course, and the present vogue is simply to leave the brush that remains after woodcutting in piles from which deer can perhaps feed during the frozen months. In the decade or more required for a brushpile to vanish into the ground, the heap of limbs can serve as a protective shelter for birds and small woods creatures. As for looks, who wants to convert a woodlot into a compulsively tidy park? Rationalizations aside, the fact remains that our times bring unexpected little deprivations embedded in the larger collective good; and for me the losses include the magic fragrance in October of leaves burning in little piles along village streets, and the bonfires that flamed in the woodlot on wet November days.

❡ OUTDOOR FIRES such as beachfires and campfires are often built with little or no surrounding structure, except as may be required for cooking or heating water. As a class they call for slightly different fire-tending protocol than do indoor ones, with somewhat modified rôles for those in attendance. There is usually still a single chief builder and tender, but outdoors he is much more likely to accept assistants and junior partners in the enterprise. Often there are volunteer aides who scour about for firewood (bringing back worthy finds with a smile of pride), branch-breakers of both the knee-snapping and thigh-smiting schools, specialists in kindling and bark, and frequently one willing fellow, more muscular than sensible, who totters back with some great timber balk more suited for battering gates than for use as fuel.

Once the fire has caught and begun to radiate, people distribute themselves about it in a semicircular pattern, fully circular if there is no wind to make the smoke a nuisance. If the hearth is ringed with rocks, and one of them is upended as a rudimentary fireback, a person wishing to talk to the group will often stand behind this rock, almost as if the hearth were a stage and the firelight the footlights. When the fire has been burning for some time, it becomes evident that the fire-builder remains the chief tender, a fire priest who determines when major pieces of wood should be added. Others around the arc have become acolytes, each concerned with policing the segment directly in front of him, pushing in burned-through pieces that have rolled free, setting alight and then blowing out the end of a poker-stick, combing the fringe of ashes. It is possible to think of this activity as no more than an absent, part-entranced playing with fire. So richly is fire interwoven with the human past, however, that it could also be the ghost of some ancient forgotten behavior from days when fire was the treasure of the band, to be served and groomed with reverence.

How people arrange themselves after dark about an outdoor fire is another fit subject for observation. There are to begin with the faint magnetic fields of people in a group, the weak but affecting force (entirely separate from pairing by sexes) that persuades us that some persons are all right and even pleasing to be near, and others are less so. (If an experimentalist is listening, I suggest he investigate a possible correlation between interpersonal spacing and the quality of voices. We may move away from people with aluminum or zinc voices.) Superimposed on this delicate influence on behavior is the much stronger effect of temperature. Entomologists report that the rate at which ants move is a direct function of temperature; the stance of people about a campfire is equally temperature-sensitive. On a soft windless summer night we often lie on the ground, not wanting to expose Sun- or wind-burned skin to infrared radiation, and bothering with a blanket or mat only if we are clothes-aware. If the night is ten degrees cooler we sit facing the fire, hands encircling knees. If it is cooler still we prefer to sit on perches of some sort, aware of the thermal absorption of the ground that is known for some reason as "damp." (The ground may or may not actually be damp, but this is not the attribute that we are trying to define.) Finally, if it is truly cold and windy—as when, on a camping trip, the blaze is a prized interlude between the trail and the sleeping bag—we stand by the fire, perhaps eating from a plate and sipping the boiled coffee that is so good outdoors and so noxious otherwise. After the Sun goes down, we do not willingly leave the blaze. As an ultimate in drawing heat from a campfire on a breezy icy dawn, we stand very close to the fire, shuffling our feet in an unrhythmic dance and rotating slowly to distribute the precious radiation, human planets by a campfire Sun.

In comfortable weather the stagey lighting created at night by a campfire is clearly part of its witchery. Indoors, we seldom rely on firelight alone, nor see the fire from a variety of perspectives nor from afar. But to approach and join a group of people around an evening campfire is to receive a sequence of powerful visual sensations. At the first glimpse of distant flames there is a flicker of alertness: is it *intentional?* Its place and constancy soon set this anxiety at rest. If you are paddling toward the fire across a lake, it enlarges very slowly, a ragged spot of orangey red, until finally you are near enough to make out tiny silent figures about the blaze, some in silhouette, and the flames make a path of reflection on the black water ahead. Unless there is a night breeze, there is a strong temptation to stop paddling and drift noiselessly toward the circle of light.

If you are walking toward the fire through the woods, it will change in brightness and apparent location as the path winds, disappearing for a time and then reappearing larger and in a slightly bent direction. After you come close enough to make out faces, some edge-lighted and others fully illumined and ruddy-looking, there comes an impulse to stop and study the scene, for you feel as invisible as a spirit, and the people about the fire seem to have been arranged by a stage director or painter. The hemisphere of light, so bright at its core, is no match for the huge mass of darkness, and can only sketch the presence of trees near by. As sparks fly upwards to the sky, there is wonder of how many generations of men have pondered if this can be the way that stars were made. When after a long moment you step forward to join the company about the fire, a dog dozing in the warmth will often be startled into barking, embarrassed by his failure to have marked your approach at a proper dogly distance. The persons around the fire are less likely to be surprised, because to gaze at bright flames is to be conscious of deep surrounding darkness.

¶ TO RETURN to the principle of concentrating heat in kindling the fireplace fire, it is obvious that one almost never wants a rapacious, fuel-controlled fire indoors. The manageable, just-critical fire that is far more desirable wants to be kindled in a positive but modest fashion, not conflagrated. (Fires, like children, are much affected by their beginnings.) You begin by carefully selecting one or two pieces of firewood around which the whole kindling effort will be concentrated. These are key pieces, no more than arm-thick, and known to be dry. They are assuredly not the backlog, which is a husky hunk of wood, not necessarily dry, that is tunked down against the rear wall of the hearth. Nor are they the front log, a slightly less husky split billet that is set on the andirons (thinner edge to the rear) an inch or two away from the backlog. It is in the slot between these two pieces—which form a miniature chimney within the hearth—that a pleasing fire can be most reliably created. Into the slot first go a few pieces of kindling, finger-thick and dry; and on top of them go the key piece or pieces, thin edge downward. If there are any charred log ends left in the hearth from a previous fire, work them in just above because their previous experience will have left them extra fire-willing. Finally the structure is capped with one or two additional pieces of firewood that will top the slot. The goal is for hot gases and flame from the kindling and key pieces to travel upward through narrow sinuous courses, and the capping pieces should not be so snug as to block this flow. A new fire needs modest internal breathing passages that an established, thermally affluent fire will create for itself. But up to limits set by a young fire's breathing capacity, the pieces want to be close together, reinforcing and reflecting each other's combustion, for fire loves proximity. Touch a match to a bit of paper beneath a fire assembled in this fashion and there is high probability of a quick and reliable blaze.

It is not argued that this method of kindling a fire has extraordinary merit and should be practiced by all apprentice tenders. There are at least a hundred ways to kindle a fire, many of them entirely workable and reliable. I think the differences are worth no more than, at best, idle and amiable dispute among friends. Accomplished tenders seem less concerned with carefully following prescribed patterns than with achieving good results with unfussy methods. (They do seem as a class to have a small vanity about using very little paper and only a couple of pieces of kindling, which is surely an acceptable form of negative ostentation.) A fireplace is in its nature unsuited to rigid dogma because every fire built in it is in some way unique. Firewood itself is not uniform (sometimes even maple can be malicious or hemlock mannerly), and the variations in size, shape, dryness, and relative position of wood are nearly infinite. This means that an experienced firetender is not so much a priest of procedure as he is an experimentalist and a close observer.

Fires reward close study, and not just in the trancelike state that is so easy to fall into. If you watch a fire with a seeing eye, you will note that throughout the fire's life there is an imperfect and unstable balance between the proximity that encourages burning and the separating effects that are created by burning. A fire is subject to at least three separating influences. The burning process, by consuming the burning surfaces, increases the distance between them. Then logs lose mass in burning, and this shifts their centers of gravity so that they tend to roll apart. Finally, logs burn through near their middle, the place of greatest averaged combustion, until ultimately they fall in tapered half-pieces. The work of a firetender is to remedy these divisive effects. With toe or poker he resettles a structure that has burned itself hollow and rickety, encouraging the wood to snuggle into warm reconciliation. His tongs fit the larger half-pieces back into the edifice of the fire, or if they are too charred and inconsequent to fit well, he puts them beneath in the bright orange coals to add their substance to the fiery excitement. Guiding his tongs during these attentions is his knowledge of how the fire will respond to proximity and reinforcement, his awareness of how superposition can cap and heat up a portion of the fire that might cool and die. Like a Merlin of the hearth he can create magical effects, as when a ragged gap-toothed blaze is delicately rebalanced into an even cascade of upward flame, flowing in a billow like water over a submerged rock.

His nearest approach to a set-piece of magic, always against the constraint that a fire is not a firework, comes when he resurrects a neglected fire. The situation commonly arises when one returns to living room or study after an absence, perhaps for a meal, to find that an untended fire has burned alone toward extinction: flames gone, blackened logs in disarray, the coals turned to apparently mortal greyness. But a knowing firetender efficiently reassembles the blackened logs on and above the coals (whose whitened look is just a mask), adds a fresh piece of wood to reflect and concentrate the heat, and sits back to wait the few seconds required for the hot gases to collect and grow hotter until, *poof*, the fire relights in bright yellow prancing flames. When I was small there were skilled firetenders in the house, and one was a maiden aunt, a gentle and self-effacing soul much concerned with Godliness. Once she told me that she possessed the remarkable power of making a fire spring to life at the snap of her fingers, and then she did it. I remember studying the hands with which she had made the magic gestures: they were not young, showing the delicate stains of age and the knuckle enlargement of arthritis. No clue could be found to the extraordinary power they had just demonstrated. Several times that winter I privately made powerful gestures and magic finger-snaps at darkened fireplaces but it never worked for me—as a result, I had to assume, of inferior Godliness.

In a brisk fire each burning log makes its individual contribution, so that the number of logs blazing roughly establishes the scale of the fire. A four- or five-log fire is the upper end of the range, useful on days of high reality levels, as when a closed house is newly opened, or when the wind after a heavy snowfall is keening at the eaves and shaking the storm sash with ugly threats. At times when you must burn unseasoned wood, a five-log fire also has merit, providing extra crevices for flame to lodge in; it gives no more heat than a smaller fire of dry wood until the time comes, after several hours of hissy semi-smoldering, when the fire has dried out its fuel and can settle down to a proper ballet of flame and coals. If it is to be a durable fire, this is of course the moment to add more green wood to carry on the drying-out process.

On most occasions, and particularly when frequent tending is inconvenient, the three-log fire is basic: three largish logs arranged in triangular profile to co-operate on the blaze. Tenders soon learn to think in terms of the age in the fire of each log. While the burning age of each log is identical at the moment of kindling, this soon changes as the pieces burn at different rates and as fresh —fire-younger—pieces are added. This in turn means that the normal three-log fire is frequently made up of one young and two middle-aged pieces. Two-log fires are more exigent than three, demanding much attention, except with extremely dry wood; and the one-log fire is an oddity, an entertainment: on occasions it is a challenge to keep a lone log aflame with bits of thermal encouragement tonged beneath it. If you are reading late in the evening, when it would be wasteful to build up a fire that would outlast wakefulness, it is possible to keep a modest, companionable flicker burning from a single log by feeding it charred ends and those morsels of wood and bark that collect in the woodbasket. A thoughtful firetender matches the kind of wood he uses with the expected duration of the fire, not using long-burning hickory or oak on a fire kindled in midevening if there is some cherry or elm at hand. If some mischance of mood and guests threatens an evening of exceptional dullness, he can add a piece of Osage orange—just one, not too big—to the fire, Osage orange being a wood of unusual pyrotechnics. It gives off showers of darting sparklets likely to arouse even the most torpid guests.

Reviving yesterday's fire when it has survived overnight but is close to extinction—a thread of smoke arising from a black lump in a field of soft grey ashes—can challenge a firetender. The setting for this is not like my aunt's resurrection play; it is not a fire gone comatose during dinner. The occasion arises when you first come downstairs in grey daylight after a pleasing evening, perhaps during a winter holiday when renewal of yesterday's fire seems an idea rich in continuity. If the house is inconveniently cold or if the brightness of daylight splinters in your eyes, it is probably sensible to discard symbolism and simply build a fresh fire, for you will get warmth faster that way. But this will give none of the satisfaction of nursing a tiny coal into new strength, kneeling to breathe on it most gently, sheltering it from sudden displacement, feeding it choice tidbits of bark and slivers, building a doll's campfire in the ashes, strengthening the first flames until they can communicate themselves to normal kindling, and finally bringing back to vigorous life a fire that has already lighted and warmed one memorable day.

THE SPIRITS OF THE HEARTH

If thousands of years of tradition are any guide, the likeliest place in your home for gods to dwell is in the fireplace. Hearth gods have been numerous beyond counting in times gone by, friendly spirits in the main, not oppressive in their divinity. Sometimes they have been earthy and a little uncouth, like Cacus and Caca, a brother and sister pair of hearth deities who had a vogue when Rome was just a tiny hamlet on Palatine Hill. If Vergil can be trusted, they were the product of a casual liaison between Vulcan and a fire-breathing she-dragon with the unique habit of using her personal flame thrower to stop, rob, and incinerate travelers. Cacus and Caca showed none of the personality damage that might have been expected of such lineage. They were plain rustic gods, concerned with the well-being of cattle as well as fire, and venerated in a rude little temple that sheltered Rome's public fire from rain and wind. Since striking new fire from flint, steel, and good dry tinder was a bothersome process, even small settlements permanently maintained a community fire from which flame could be taken, as a kind of reverse fire department. Cacus and Caca presided over an informal temple, a sociable place for farmers, herdsmen, and townsmen to drop in for a light.

But as Rome grew and persevered on the course that led toward grandeur, barnyard deities were unsuitable and Vesta became a preferred hearth goddess. (There is just a possibility that Vesta might have derived from a tidied-up version of Caca, a scrubbed Caca sent to charm school; but most classic scholars believe she developed from Hestia, a Greek goddess of impeccably goddessy attributes.) Vesta was a queenly deity with no sideline responsibilities over cattle, although she did have a fondness for bakers and millers, and her favorite animal was the ass, the beast that turned millers' grindstones. In homes she was present in the domestic hearth and was formally reverenced three times a month with wreaths provided by the chief female house servant. In her public municipal temple Vesta reflected both a kind of early feminism and a Roman weakness for progressively elaborated ceremony. The Vestal virgins seem to have begun informally, a few young girls brought in for firetending chores and sweeping out the temple. In a few centuries Vesta and her virgins had become a powerful, rule-encrusted institution, with permanent staff, money, authority, and an impressive temple. Political pull was the first of many qualifications for admission to the corps of virgins. The trend may be surmised by the punishment structure: a virgin who allowed the sacred fire to go out received a beating, but one who lost her chastity was ceremonially buried alive. No breath of scandal ever touched Vesta herself, although she was known to have a genteel relationship with Fornax, the tend-to-business god who saw to bakers' ovens.

The native Roman bent was reflected less in the imaginative creation of fire gods than in the artful management of fire itself, notably the fires that warmed the elaborate public baths. The Hellenistic contribution to the *thermae* was relatively minor; in the 5th Century B.C. Herodotus described a Greek custom of heating steam baths by dropping fire-heated stones in water, a practical technique but one already used by many peoples. The special Roman talents for masonry, hydraulics, fire management, and large-scale organization came together to create baths that would take platoons of engineers to duplicate today. They began with great vaulted *tepidaria,* which were gently heated spacious rooms adjoining exercise courts and sometimes fitted with glass-walled alcoves for Sun-bathing. Then came the *calderia,* hotter and more humid, that induced profuse sweating; *laconica,* rooms of intense, barely tolerable dry heat; and finally *frigidaria,* with plunging pools having different degrees of coolness. The effects were achieved by leading under floors and through hollow walls the flue-gases from wood and charcoal fires, and by lavish use of hot and cold water. An intricate network of masonry and terra cotta ducts was employed. The *thermae* were institutions without equivalent today, although our country clubs are a faint echo, a major difference being that the baths usually were open to all freemen and charged only a token admission fee, since they were state-subsidized. They offered, besides exercise and baths, massage and a kind of chiropractic joint-cracking, wine and delicacies, erotic entertainments, a chance to learn the latest military and political news, and occasional chances for personal preferment. It amounted in all to a richly varied way to spend an afternoon.

In provincial cities baths were also common if less elaborate. On the outposts of empire north of the Alps, where winters were harsher, many hypocaust-heated buildings were constructed for officials, and similar arrangements were used in villas built for the rich. These were structures where flue-gases from fires were led through the floors and walls, giving a radiant central-heating system that would not be matched technically for almost two thousand years. In lesser buildings for lesser ranks, it was the Roman custom to use portable charcoal braziers to create local circles of warmth.

As the empire broke up and its cultural fabric tore apart, its technology disappeared from men's memories. We do not always show a full understanding of the processes at work in the decline of a high civilization. As a consequence it can surprise us that in the rough fortress-castles built during the Dark Ages, a fireplace was sometimes literally a place for fire, without chimney and with only a gap in the roof to let some smoke out—a stone equivalent of an aboriginal hut. It was not that the hand of man had inexplicably lost its cunning, nor that the amenities were brutishly devalued. What happened was infinitely sadder. After century upon century of savagery, pillage, and murder, preoccupation with survival so overwhelmingly dominated all other things that it was not just technique that had become forgotten but also the existence of anything other than survival itself, a preoccupation that ruthlessly effaced such concepts as comfort, ease, and civility.

¶ THE FIREPLACE that we know, essentially an open-sided recess topped by a chimney, was built in Europe from early mediaeval times. During the Renaissance, the fireplaces in great halls and palaces grew increasingly grandiose as skilled masons and bricklayers became available to indulge the tastes of the rich and powerful. Giant fireplaces were much favored as symbols of conspicuous consumption long before yachts and similar flauntings had their vogue: a personage whose blaze was fed by eight-foot logs borne in by a pair of servants was obviously no mean fellow. The impressive theatrics of such a fireplace may have outweighed its serious drawbacks, which were a monstrous appetite for wood, an unstable zone of comfort, a tendency to smoke on an epic scale, and a disposition toward extreme draftiness as large volumes of air were drawn up the chimney. It was not the first demonstration of the fact that it can be uncomfortable to be impressive.

Early fireplace masons evidently believed that the remedy for smokiness was to build a chimney having a larger cross section, and to deepen the fireplace itself. Both were measures that only slightly palliated the problem, and at a very considerable cost in efficiency. Surviving 17th Century manorhouses often have fireplaces of proportions that would very nearly suffice for a bull-roast, and these deep caverns must have given very poor return for the wood they swallowed up. It was not until the 18th Century that it slowly became recognized that in fireplaces the biggest is not necessarily the best, and that a simple scaling-up of proportions toward the grandiose was a prescription for trouble.

Benjamin Franklin, whose exceptional mind brought him distinction as a printer, publisher, author, inventor, businessman, scientist, statesman, and diplomat, was fascinated all his long life by the behavior of fireplaces. In his *Autobiography* he wrote:

> I should have mentioned before that having, in 1742, invented an open stove for the better warming of rooms, and at the same time saving of fuel, as the fresh air admitted was warmed on entering, I made a present of the model to Mr. Robert Grace, one of my early friends, who, having an iron furnace, found the casting of the plates of these stoves a profitable thing, as they were growing in demand . . . Governor Thomas was so pleased with the construction of this stove . . . that he offered to give me a patent for the sole vending of them for a term of years; but I declined it from a principle which has ever weighed with me on such occasions, viz., that, as we enjoy great advantage from the inventions of others, we should be glad of an opportunity to serve others by any invention of ours; and this we should do freely and generously.

Franklin can easily be forgiven the slip in misrecalling the date—the stove was invented in 1740, and advertised in his own newspaper in 1741—and for not recalling that the first models were cranky beasts, suffering from being overcomplex, that common ailment of early inventions. The first ones were like modern heating fireplaces that circulate hot air through side vents, but it wasn't until the design was somewhat simplified that the stoves, which Franklin called Pennsylvanian fireplaces, worked reliably. In a sales pamphlet he wrote for Robert Grace in 1744, Franklin revealed that shrewd copywriting was another of his skills. He described the dangers of ordinary fireplaces that created drafts of cold air on the backs of those attempting to warm themselves: "it rushes in at every crevice so strongly as to make a continual whistling or howling; and 'tis very uncomfortable to sit against any such crevice . . ." He quoted what he described as a Spanish proverb:

> If the Wind blows on you thru' a Hole,
> Make your Will and take care of your Soul.

Then, showing that he realized the influence of women on domestic purchases, he aimed directly at them: "Women, particularly, from this cause (as they sit much in the house) get colds in the head, rheums, and defluxions, which fall into their jaws and gums, and have destroyed early many a fine set of teeth in these northern colonies. Great and bright fires do also very much contribute to damage the eyes, dry and shrivel the skin, and bring on early the appearance of old age." Closed metal fireplaces (stoves) were fairly efficient for heating, he admitted, but because people, the English particularly, loved an open fire, stoves tended to be neglected and were often too hot or too cold. But with his stove-fireplace comfort was assured, and people could sit by a window in good health to sew or read.

Franklin was not a zealot, and did not reject conventional fireplaces. In 1758, in England on the first of his prolonged, delicate, and generally successful diplomatic missions, he wrote a letter to someone called "J. B., Esq., of Boston" in which he invented the concept of a movable iron plate in the chimney throat—in other words, a damper—as a means of adjusting the draft according to need. It is a letter that affords a glimpse into the workings of Franklin's mind. Challenged by the saying "as useless as a chimney in summer," he immediately concluded that chimneys need *not* be useless in summer. Because temperatures inside and outside the house were almost never identical, drafts existed there that could be put to work. "If the opening be closed by a slight moveable frame or two that will let air through, but keep out flies, and another little frame set within the hearth, with hooks on which to hang joints of meat, fowls, etc., wrapt well in wet linen cloths, I am confident that if the linen is kept wet, by sprinkling it once a day, the meat would be so cooled by the evaporation, carried on continually by means of passing air, that it would keep a week or more in the hottest weather. Butter and milk might likewise be kept cool, in vessels or bottles covered with wet cloths." This notion, which is technically entirely sound, was tossed off by a man immersed at the time in thorny and often hostile negotiation with the proprietors of the Pennsylvania colony.

Much later, in fact after Franklin's death in 1790,

there came to light a copy of a long letter he had written to a friend, the physician to the imperial Austrian court. It was written during the enforced leisure of shipboard on Franklin's last Atlantic crossing in 1785, returning to serve in the Constitutional Convention, as its oldest member. "The garrulity of an old man has got hold of me," he wrote his friend, "and as I may never have another occasion of writing on this subject, I think I may as well now." He was in fact eighty-one, and bright as a button. After his death the letter was converted into a treatise and published in 1793 under the title *Observations on Smoky Chimneys, Their Causes and Cure.*

He made short work of contemporary theories that a chimney should have an enlarged internal diameter as it rose, and a converse theory that its internal dimensions should gradually decrease. Within extremes, neither shape made any difference. He had conducted simple experiments with bits of silken thread used to detect movement of air in glass tubes of different shapes. The shapes didn't matter; the only condition that moved the threads was air heated in the tube. "No form of the funnel of a chimney has any share in its operation or effect respecting smoke, except its height."

Franklin then ticked off causes of trouble. "Smoky chimneys in a new house are such frequently from mere want of air . . . When you find, on trial, that opening a door or window enables the chimney to carry up all the smoke, you may be sure that want of air *from without* was the cause of its smoking." Air from other parts of the house was not the answer; outside air was needed.

Simply leaving the door or window open a little was unsatisfactory because it created cold drafts on the back and feet of those by the fire. Since air up by the ceiling was warmest, a better plan was to cut an opening high on an opposite wall so that cold incoming air would be mixed with the warm. Briskly Franklin explained that the necessary size of the opening could be determined by gradually closing an outside door to the point where the smoke just reliably went up the chimney, and then measuring the crack. If a six-foot door needed to be a half-inch open, an opening six inches square should suffice. An opening high on a wall could be concealed by putting a shelf just beneath it. In England he had noted cases where an upper pane of a window had been replaced with a disk of rotary metal vanes; the idea was sound but it was annoyingly noisy and could be improved.

The old man's agile mind danced on to another common cause of smoking fireplaces: "their openings into the room were too large." This often occurred because architects were more concerned with the looks of things than with performance. The size of a fireplace opening ought to be in proportion to the height of a chimney. The bigger the opening the taller the chimney should be. It followed that fireplaces on the second story should be smaller than those on a ground floor. Overlarge openings were unsatisfactory because they admitted unheated air to the flue, weakening the draft that carried smoke away. Oversmall openings, although not common, were undesirable because they created too strong a draft, which consumed wood with wasteful speed. If owners and architects insisted on an appearance of grandness, let them use marble facings about the fireplace to give an impression of largeness. Normally an equable and even genial man, Franklin could be testy with folly: "And [yet] there are some, I know, so bigotted to the fancy of a large noble opening that, rather than change it, they would submit to have damaged furniture, sore eyes, and skins almost smoked to bacon." It is possible to infer that somewhere on his busy travels, the famous American statesman had prescribed a remedy for a miscreant fireplace that, from pride and vanity, had gone unused.

His letter went on in the abundant practical detail that reflected a lifetime of observation and curiosity. Too short a chimney was another frequent cause of smoking. The interrelationship between fireplaces within a house needed study to "prevent their overpowering one another." A powerful fireplace could reverse the draft in a weaker one, and for this an ample supply of outside air was the answer, as well as sliding metal plates (dampers) within chimneys to regulate the draft. Other problems were examined—the need to proportion chimneys to permit passage by chimneysweeps, the special problems of kitchen fireplaces, what to do when a door was unfortunately located too close to a fireplace, and three remedies for circumstances when "the tops of chimneys are commanded by higher buildings or by a hill." Before he ended, Franklin had covered nine major causes of smoking chimneys, the remedies for each that he had found of value, as well as various special cases he had seen and how *they'd* been remedied too.

The plump, bald old man scribbling away so industriously in his ship's cabin was plainly an acute observer of fireplaces and a gifted experimenter. Fifty years before in his *Gazette* he had published a letter he had written himself, purporting to be from an old man concerned about the control of accidental fire, "the fiercest enemy of property." The letter cautioned against carelessness in carrying hot coals about, and then went on to his real purpose, which was to recommend the forming of a society of men in Philadelphia experienced in putting out accidental fires once started. Forty-five years earlier he had invented the Franklin stove; and only fourteen years ago he had invented in London a coal stove that in some degree consumed its own smoke. He was the inventor of lightning rods to save men's houses, barns, and ships from capricious destruction, causing Immanuel Kant to proclaim him a new Prometheus who had stolen heaven's fire. Yet fire was only one of his interests. He was an electrical pioneer, the first importer of silkworms and osier for wicker; an advocate of public libraries and insurance companies, an enthusiast for such diverse ideas as weather forecasting, polar exploration, and lime fertilizer; a widely read author and aphorist; the most unfailingly adroit diplomat in America; and the only man among the great ones of his time whose signature appears on all four historic documents of the day, the Declaration of Independence, the alliance with

France, the peace treaty ending the revolution, and the Constitution.

¶ IN 1796, just three years after Franklin's letter-turned-treatise appeared, an important paper on the theory and practice of fireplace construction was published in London. It was called *Of Chimney Fire-places, with Proposals to Save Fuel; to Render Dwelling-houses more Comfortable and Salubrious, and Effectually Prevent Chimnies from Smoking.* Its author was Count Rumford, a handsome, dashing, and thoroughly self-assured man identified on the title page as Knight of the Order of the White Eagle, and of St. Stanislaus; Chamberlain, Privy Counsellor of State, Lieutenant-General in the Service of His Most Serene Highness, the Elector Palatine, Reigning Duke of Bavaria; Colonel of his Regiment of Artillary; Commander in Chief of the General Staff of His Army; F.R.S. . . . (the identification continues at further length). The Count wrote that he had overseen the altering of five hundred faulty fireplaces, of which all had achieved immediate improvement except one that had needed some trifling extra attention. His methods, he reported, would save more than half the amount of fuel previously consumed, and were equally adapted to fireplaces burning wood, coal, or peat. They eliminated dangerous and indeed often fatal drafts, and rarely needed the admission of cold outside air. The only objections he recalled receiving were from persons who said that his fireplaces rendered rooms too warm;

but pity was the only proper response for persons lacking the wit to know how much fuel to put on a fire.

An adventurous and versatile figure, Count Rumford had among many other occupations a rôle as a fireplace doctor with a large society practice in England and on the Continent. The prescriptions set forth in his essay were almost precisely those that an experienced engineer might use today to improve the wasteful cavernous fireplaces that had been traditional for centuries in manor houses and stately homes.

To begin with, the throat of the fireplace (the transition from fireplace to chimney up behind the lintel) was narrowed to a slot not more than four inches wide, reducing the amount of heat lost up the chimney. This narrowing was done by building a flat-topped smoke shelf on the back chimney wall, which additionally served as a deflector of any downward puffs of air occurring when only a small blaze was burning. Rumford was the first fireplace theoretician to recognize that in a small-blaze situation there might be two-way air flow within a chimney, with cold air descending by the back wall while hot air and smoke rose by the front. Also, the back of the hearth was filled in with masonry until the fireplace was much shallower, with the fireback curving toward the room as it rose, for better heat reflectance. The sides—then called jambs or covings—were modified from their old boxy configuration until they flared outward, also for better infrared reflectance. The width and height of the fireplace opening were not allowed to follow some departed builder's whim but were regulated by a strict set of proportions. Because not even the most rachitic and malnourished chimneysweep could be expected to squeeze through a four-inch slot, Count Rumford specified that a few bricks or stones at the top of the fireback should be installed without mortar so they could be removed to permit the lad's passage, and later put back to restore the fireplace to thermodynamic perfection. He preferred brick or stone as materials with which to fill in oversize fireplaces rather than iron, for "iron, and, in general metals of all kinds, which are well-known to grow very hot when exposed to the rays projected by burning fuels, are reckoned among the very worst materials that it is possible to employ in the construction of Fire-places." In similar tones of cocksure pedantry he specified that the fireback and walls should be made as smooth as possible. Indeed for an absolute maximum of heat and light, these surfaces should not be allowed to grow black and sooty but should be frequently cleaned and whitewashed.

It must be said that, disregarding the Count's unwillingness to bow to the rituals of conventional modesty, his fireplaces were superb, and not just in comparison to the cranky and wasteful ones they replaced. Some survive today (both modified ones and those built from scratch); and with the addition of an adjustable damper are notably efficient performers. A fireplace built to the Count's proportions seems a little shallow to modern eyes, but this is because we have become accustomed to the deep rectilinear grottoes created by architects in their quest for a "focus of interest" along one wall. It is our taste that may be impaired, for in heat delivered and disinclination to smoke the Rumford fireplaces are among the most sophisticated that have ever been built.

Rumford was in London in 1796 to see to the publication of a multivolume work of his writings, of which the fireplace essay was part. His stay was cut short by a courier from his patron, Charles Theodore, the Elector of Bavaria. Hostile Austrian and French armies were marching against Munich. The Count was to return at all speed to lead a Council of Regency for Bavaria, since the Elector deemed it prudent to go to cover elsewhere.

At this juncture it is worth examining the career of the remarkable Count. A portrait of him by Gainsborough reveals a strikingly handsome man with an expression that combines quick intelligence and impatience or arrogance. He had been born Benjamin Thompson, on a small farm in Woburn, Massachusetts, in 1753. His farmer father had died early, and he had been apprenticed at thirteen to a Salem shopkeeper. He studied medicine briefly, and taught himself astronomy and mathematics. Although unregistered at Harvard, he had sat in on the lectures of John Winthrop, professor of mathematics and natural philosophy, whom the precocious ex-farm boy characterized as "that happy teacher." Another facet of his precocity was displayed when, at nineteen, newly appointed as schoolteacher at Rumford, New Hampshire, he married a rich and well-connected widow fourteen years his senior, who had fallen in love with him at sight.

This success in no way blunted his ambition. In fact on his honeymoon at Portsmouth he snatched at an opportunity to impress the royal governor, appearing at the annual military muster on a restive white stallion, dressed in a scarlet jacket surmounted by a blue hussar cloak with "mock-spangle metal buttons." Governor John Wentworth invited the couple for dinner and soon became a patron of the remarkable Yankee youth, so handsome and charming, so formidably well informed.

But the troubled times intervened. Thompson was a Loyalist, disdaining the local patriot-rebels, and his arrogance poisoned his relationship with the town. In 1775 he was haled before a local Committee of Safety, charged with sending secret intelligence to General Gage in Boston; the evidence was inconclusive and he was acquitted. Local distrust remained so deep that he had to leave town, slipping quietly away from the mansion where his wife and infant daughter remained. He went to his native Woburn, and there soon the pattern repeated. Evidence surfaced after the second trial that he had in fact been sending information to Gage with invisible ink at the same time that he was indignantly protesting his innocence (but the difference between intelligence work and treason is viewpoint). After his second acquittal the risk of tar-and-feathers remained so great that he fled town once more, ending up in Boston within British lines. Here he presented Gage with a detailed report titled "Miscellanius Observation upon the State of the Rebel Army."

Thompson made his way to London, where his report

so impressed the Government that Lord Germain, the colonial secretary, offered him a post in the Colonial Office. Like the New Hampshire governor before him, Lord Germain was much taken with the brilliant Yankee youth and made him his protégé. Envious London rumors put the annual income from his various posts as approaching £7,000. His energy and versatility had no limits: he also found time to conduct scientific experiments on heat and gunnery, become a fellow of the Royal Society, keep a sequence of desirable mistresses, become a fashionable doctor of faulty fireplaces, and devise technical improvements aboard H.M.S. *Victory*. These were no small achievements for "the Woburn lad," a provincial still in his twenties, and testify to his intelligence and exceptional gift for self-advancement.

Another of Thompson's talents, not altogether surprisingly, was making enemies. Before long, London gossip linked him with a newly captured French spy, and once again the young man found it prudent to resign his posts and leave town. He soldiered for a while in the Carolinas, and Light-Horse Harry Lee wrote after the war of one engagement when Thompson's dragoons surprised and mauled a larger American force. In winter quarters in Huntington, Long Island, in 1782–3, Thompson stabled the mounts of his regiment in the local church and set up camp in the adjoining cemetery. The townspeople were particularly inflamed when regimental cooks baked bread in tombstone-floored ovens, so that parts of grieving inscriptions were reproduced in relief on the bottoms of regimental loaves. No one in Huntington that winter found the twenty-nine-year-old lieutenant-colonel the least bit charming.

Thompson, no man to soldier to the bitter end of a lost war, made his way to London, received a promotion to full colonel, and resigned with a half-pay pension for life, announcing his intention to become a soldier of fortune. He made a channel crossing in the company of historian Edward Gibbon, who addressed him with indeterminate irony as "Mr. Secretary, Colonel, Admiral, Philosopher Thompson." In Munich he so charmed Charles Theodore, the Elector of Bavaria, that he was asked to join the Elector's service. Thompson said that he would have to return briefly to England to obtain his sovereign's permission. It was a curious specification in the light of George III's personal dislike of Thompson. But as always in such matters Benjamin Thompson knew exactly what he was doing: he emerged from Government conferences with permission to serve the Elector, and with word that he was to be knighted. Technically this was to give him appropriate rank, but there seems little question that an intelligence relationship was also constructed, with knighthood to seal the bargain.

In Bavaria Thompson showed skill as a reformer and administrator, and spent enormous energy in correcting the dishevelments he found in that country's affairs. Munich's streets were clotted with beggars. He had twenty-six hundred of them scooped up in a single day, fed, clothed, sheltered, given instruction in making shoes and clothes for the army, and paid wages for it. "To make vicious and abandoned people happy," he observed, "it has generally been supposed necessary first to make them virtuous. But why not reverse the order? Why not first make them happy, and then virtuous?" He reformed the army, raised pay, improved rations and quarters, gave soldiers regular home-leave, founded training schools, encouraged sports competitions, built up the artillery into an elite group, and gave each soldier his own garden plot. Landscape gardening caught his fancy, and Munich's Englische Garten was designed and created by Thompson, assisted by an army corps to drain away the swamps and otherwise revise the landscape. In 1792, when authority devolved on Charles Theodore during an interlude between emperors, the grateful patron honored Thompson by naming him a Count of the Holy Roman Empire. For his title the new Graf selected the little New Hampshire town where he had had his first success, finding no irony that it was in a country where he would have been less than welcome.

When Rumford devoted his intelligence and imagination toward science, there was nothing stagey or grandiose about the result: he was cautious and thorough, a first-rate scientist. In all he wrote more than seventy scientific papers, most of them dealing with heat. He was a powerful advocate of the theory that heat was energy, or molecular motion, and he attacked the entrenched caloricists with a number of experiments demonstrating the inconsistencies of their views. He was a compulsive inventor, a man who saw room for improvement wherever he looked. He invented a new carriage wheel and a method of steam-heating houses, a shadow photometer and an accurate calorimeter, the use of preprinted petitions to speed the business of state, a variety of clever cookstoves for Bavarian soldiers, a greatly improved oil lamp, a roaster and pressure-cooker, and teakettles that look singularly modern today. Discovering coffee, he immediately wrote a long essay in praise of its excellent qualities. Characteristically, he thereupon invented the drip coffeemaker and several ingenious coffee cups. At one period he dressed only in white: it was thermodynamically the most efficient clothing.

In Bavaria as elsewhere, however, Rumford had sown a rich crop of personal enmity, which grew in time to the point where the Elector was forced to get rid of him. Charles Theodore attempted to soften the blow by appointing him Minister to England. This was quite unacceptable to the Foreign Office, possibly because of the unconscious effrontery of having its own agent posted back to His Majesty's Government. Rumford nevertheless was admitted to the country. In England he let American officials know of his willingness to be appointed the first superintendent of the new U.S. Military Academy, then being formed. This poignant request was for a time considered until someone in America realized its total inappropriateness. In 1799 he began devoting his energies to organizing with Sir Joseph Banks the Royal Institution, an influential precursor of the modern museum of science, and he obtained the services of Humphry Davy as first lecturer. But in 1802 Rumford had a sharp falling-out with his associates, and angrily left England forever. He lived out his last dozen years at Napoleon's sufferance in a Paris suburb, energetically conducting experiments on heat. He died in 1814 at the age of sixty-one, leaving his military library to West Point and the bulk of his estate to Harvard. It was to found a new professorship "to teach by regular courses of academical and public Lectures, accompanied by proper experiments, the utility of the physical and mathematical sciences for the improvement of the useful arts, and for the extension of the industry, prosperity, happiness, and well-being of society."

The thermodynamically elegant Rumford fireplaces, reflecting radiant heat from their smooth outward-flaring sides and curving firebacks, poured out a higher proportion of heat to the room than did any previous fireplaces. By the standards of their times, and perhaps even by ours, they appeared daringly shallow, with the fire seeming to be almost in the room, and yet they were generally smoke free. But contrary to Emerson's better-mousetrap doctrine, Rumford's proportions did not spread immediately all over the world. A few venturesome people everywhere built them, and some still do,* but they were never absorbed in the main current of fireplace design.

* I know well an unpretentious little fieldstone fireplace in Maine that by accident closely reproduces the Rumford proportions. It was built decades ago by a mason-carpenter-trader-farmer, a moody, gifted man who surely never heard of the Count, and who had a reputation for stubbornly building things not so much according to plan or request as to his mood of the moment. Years later he chanced to come in one rainy autumn afternoon to set a spell. "Almost forgot that fireplace," he said, eyeing his creation with curiosity. "Ugly bastid. Works good, though, don't it?"

It is hard to be sure why. One partial explanation is that they called for a little extra masonry skill. Another might be that they appeared at a time when the metal stove was beginning to take over the space-heating part of the fireplace's function, and thus never became assimilated in masonry tradition. Perhaps the likeliest reason is that Rumford's design looked unconventional, and fireplace masons are in general conservative men. They recognize that a fireplace may last a long time, long after those around now are dead and gone; and masonry is costly and difficult to rebuild. Since no mason wants to have his name associated with a smoker, it is obviously prudent—just to be on the safe side—to make the fireplace deeper and boxier than is probably necessary. The tendency continues today, almost two centuries after the Count showed otherwise, in architectural handbooks and recommended Government designs. Prudence is, after all, defined as a virtue, even if wrong.

¶ AS THE Iron Age came to its 19th Century climax, masonry fireplaces lost their previous dominance, slowly giving way to stoves and to iron fireplaces. In the first half of the century iron fireplaces had a great vogue, growing to be possessions of great pride and, it was felt, beauty. They could be had with iron stars, pyramids, palms, ferns, fronds, rosettes, dadoes, fretwork, urns, classical females, scrolls, improving mottoes, spears, arrows, cherubs, eagles, stallions, and other varieties of tasteful ornamentation. One can see the intersection of interest that made them so popular: to the inventor and patent-holder they represented perfection at last in efficient heating and the prevention of perilous drafts; to its owner the iron fireplace represented efficiency, beauty, and visible prosperity; and to the ironmonger they were products that sold well and didn't deteriorate in stock.

The wood or coal stove for space-heating came into broad use for somewhat different reasons. Its way was paved in the 18th Century by the cooking stove; in a keeping-room a cooking stove was infinitely more convenient and practical than a cooking fireplace, and it demonstrably made its room the most comfortable one in the house. When used for space-heating, a stove did not have the light and visual charm of an open fire, but it was out in the room with a large radiating surface, and it gave a fine return in warmth for the fuel burned. Moreover it was wonderfully controllable with its drafts and chimney damper, so that it could be managed easily and banked for overnight warmth. Soon stoves appeared in an immense variety of sizes and forms, many of them reasonably priced; and, with low-cost stovepipe, they were simple and inexpensive to install even where no chimney was within reach.

It is significant that, despite all the sensible and convenient advantages of heating stoves, the cheerfully blazing fireplace never went out of style. A fireplace filled some need that a stove did not; and if a man prospered he acquired a fireplace. The new rich of the late 19th Century, responding to some baronial instinct long before Matthew Josephson popularized the concept, dis-

played a notable predilection for fireplaces in the lavish mansions constructed to enhance their esteem. In fact the hunger for fireplaces in their mansions was so great as to make a problem for architects. A certain profusion of chimneys was desirable, but one hardly wanted a hedgehog of them against the sky. Since the rule that every fire needed its own flue was inviolable, the solution was to bunch the flues by sixes and eights in individual chimneys, which was a nice way out of the difficulty. Something of this same inclination for fireplace self-indulgence was shown by Mark Twain when he built his home in Hartford, Connecticut. Directly above the dining-room fireplace, he specified, was to be a large window, with the flue split and diverted to each side. He said he liked to look at the flames, and to watch snowflakes falling outside, and would enjoy doing both at the same time. This was perhaps only slightly more eccentric than the caprice of Richard Nixon, who was said to have had the air-conditioning equipment in one lodge at Camp David made more powerful, so the pleasures of a leaping fire could be enjoyed on even the most sweltering summer day.

It can be observed in our time that, particularly when decorators and architects have been allowed a slack rein, a feverish quest for fireplace novelty manifests itself. Ell-shaped and two- and three-sided fireplaces have had a vogue. There has been a rash of raised hearths, culminating in rectilinear fireplaces, starkly plain and glass-enclosed, that are about as cozy as a postoffice box. The desperate search for difference has also created the noback no-sides fireplace, a kind of indoor campfire surmounted by a metal smoke hood. Magazines vending home-decorating ideas have portrayed so many bizarre fireplaces that they convey the unintended impression of hordes of persons restlessly searching for some new arrangement that is smart, different, and that *can be talked about*, as though conversation were a problem, as indeed it may be.*

Some old fireplace concepts have popped up in new guises. Franklin's Pennsylvania fireplace has been reborn in a variety of free-standing metal fireplaces of contemporary design, vented by stovepipes and often of pleasing looks and performance. Usually they do not accept very husky billets of wood and thus do not lend themselves to great durable blazes, and they generally lack the attribute of fieldstone fireplaces of providing a choice of places to cock your feet up in comfortable disarray. Nevertheless they are practical, reasonably priced, and infinitely superior to no fireplace at all. A second old idea become active again is the air-heating fireplace —a welded steel fireplace core that, besides providing normal infrared fireplace radiation, also supplies hot air to side vents or to ducts leading to another room. Some are available with fan-forced hot-air delivery. These prefabricated cores are attractive in circumstances when a

* The following classified advertisement, from the *Washington Post* in 1973, is instructive:

> FIREPLACE—18th C w/blt-in radio,
> concrete logs, filled with plants
> & exquisite brass fender. A real
> conversation piece. $550.

fireplace is used to reduce the length of a central-heating season, and when provision has to be made to keep a home warm during winter power interruptions. I have lived with several such prefabricated-core fireplaces, and have found them agreeable. While their proportions tend toward a depth that would have risked Rumford's contempt, they behave civilly and are perhaps a better risk than putting oneself in the hands of a mason of unknown skill. (And perhaps sensibly so: we hardly expect plumbers to shape and fire toilet bowls on site.) The only vagary I have observed in such fireplaces was the tendency of one, possibly from age or a quirk of installation, to give forth, sometimes, a deep metallic boom as it warmed up. This oil-canning, as metalworkers call it, was an idiosyncrasy of this particular fireplace. I got used to it but the dog, an inveterate sleeper-before-the-fire, always leapt straight up in the air and came down running.

THE CRYING OF THE WINTER WIND

TODAY we think of fire mainly as a source of warmth and light, a comforting amenity. But from the beginning of recorded custom, men have also used fire for a quite different purpose: the invocation and propitiation of unseen forces. The candles that have been much entwined with religions were not simply to illuminate altars and temples. It is surprising to see how much smoke drifts across the pages of *The Golden Bough*. Fraser's great compilation of folkways shows the remarkable linkages between local customs, magico-religious practices, and past faiths. He writes that many scores of fire festivals have been celebrated by simple peoples all over the world, and "the mountains are ablaze with bonfires on Midsummer Eve . . . and from the valleys and the plains answering fires twinkle through the gathering gloom." Sir James was writing of Silesia, but it could also have been Moravia or Bohemia, Brazil or Bolivia, Denmark or Norway, France or Spain. Although many bonfires were believed to celebrate the day of a locally esteemed saint, or Christmas, or the end of Lent, it was not difficult for Sir James to show close linkages with winter and summer solstices, or the mileposts of the agricultural year, or with predecessor faiths. The blazes of Saturnalia presaged the Yule log, and the curious Beltane fires of Scotland were traceable to seemingly obliterated Druid customs. The annual extinction of all old fires in a community, formally replacing them with ritually kindled new flame, occurred on such a world-wide scale that it reflected a universal perception of some sacredness to household fire.

At first the details of fire festivals seem engaging, a country-fair frisking by isolated people whose lives must otherwise have had scant variety and color. When the unmarried young must take turns jumping across a bonfire, it appears little more than a preliminary courtship rite, a Jack-be-nimble variant on bobbing for apples. If wooden discs are set afire at dusk and then hurled high in the air, or if worn twig-brooms are daubed with pitch, set alight, and then whirled in the darkness so fast they make a circle of fire, it doesn't seem so different from youngsters' gamboling on the lawn with sparklers on a summer evening. When the men of the community rolled a cartwheel to the top of a hill, set it on fire, and then sent it careering downhill through the fields, leaping and bouncing and shedding sparks, it can seem a primitive analogue of the fireworks at the American Legion grounds on the Fourth of July.

But Fraser invited a more penetrating look. Jumping across bonfires might be courtship sport, with the competitive young men countered by the pulled-up skirts of the young women; but it could also have been a remnant of that grimmest of all courts, trial by fire. Flaming discs, twirled besoms, and blazing cartwheels could be a means—the only known means—of driving off the malign spirits that brought pestilence to livestock and men, and that could wither whole crops overnight. Sir James records that cats and snakes, the familiars of witches, were often thrust in woven wicker cages and immolated on ritual bonfires. It was known all over the world that a cooled, blackened ember taken from a magic fire and secured in the thatch atop a house or barn would avert lightning bolts for as long as a year, after which the old ember could be replaced with a magically potent ember from the new fire. And if the bonfires of the fire festivals all over the world were no more than rustic highjinks, why was a human effigy—long before Guy Fawkes—so often consumed in the leaping flames?

The concept of ritual human sacrifice is so abhorrent to us today that we don't grasp the logic that it once seemed to have, and so we put it down to a peculiarly revolting barbarism. But if one believes deeply in a theology that postulates dark spirits fully capable of meting out the worst misfortunes unless appeased, sacrifice becomes rational, the action of a prudent man. Friar Bernardino de Sahagún, a Spanish missionary who recorded Aztec customs as a preliminary to conversion, left this account:

> Behold what was done when the years were bound, when was reached the time when they were to draw the new fire. First they put out fires everywhere in the country round. And the statues, hewn in either wood or stone, kept in each man's home and regarded as gods, were all cast into the water. Also were these cast

> away—the pestles and the three hearth stones; and everywhere there was much sweeping. Rubbish was thrown out; none lay in any of the houses.
>
> And when they drew the new fire, they drew it there at Uixachtlan, at midnight, when the night divided in half. They drew it upon the breast of a captive, and it was a well-born one on whose breast the priest bored the fire drill. And when a little fire fell, when it took flame, then speedily the priest slashed open the breast of the captive, seized his heart, and quickly cast it there into the fire.

As if to show that the Aztecs were not alone in barbarism, the Spaniards imposed their rule with such savagery that in parts of Nicaragua, for example, it took four hundred years for the population to regain its pre-Conquest level. Nor should it be thought that blood-curdling savagery was solely a New World custom. Consider young Isaac's childish question to his formidable father: "Behold the fire and the wood, but where is the lambe for a burnt offring?" No one's rôle in this tense drama recounted in the twenty-second chapter of Genesis is free of perplexity. Was Isaac not only innocent but perhaps a little slow-witted? How did Abraham, a prophet gone grey in the service of the Lord, manage to suppress even a syllable of protest at Jehovah's sadistic commands? And how can we charitably interpret a Lord Who is first suspicious, then vengeful, and finally mollified by the sacrifice of a stray ram so doddery as to get his horns caught in the bushes?

The prophet who did the most for fire, and whose gentle doctrines did much for the uncompleted task of civilizing men, was Zoroaster. (His name was Zarathushtra in the original Persian, but comes down to us in a corrupt Greek form.) He was an historical figure of the 6th Century B.C., an Iranian of good family, moderate wealth, and luminous intelligence who just barely escaped deification, the normal fate of founders of successful religions. Zoroaster reformed the ancient polytheism of Iran—a dark field of harsh and vengeful gods, imperfectly propitiated by blood sacrifice and drug-assisted rites—replacing it with a simpler and less troubling monotheism. His deity was Ahura Mazda, the Wise Lord, who is surrounded by six sons or entities representing such facets of creation as Good Mind, Beneficent Devotion, Wholesomeness, and Immortality. Preeminent among these son-entities is Truth, who presides over fire, the universal and eternal principle. Zoroastrian ritual rejected blood sacrifice, replacing it with reverent rites before a permanently blazing fire. Zoroaster even dealt gently with evil—a challenge for all architects of new religions—with the humanist view that even the wicked ultimately achieve paradise, after an interval in hell, and that the greatest of all sins is a lie.

Aided by the military triumphs of Cyrus and Darius, Zoroastrianism spread until it was dammed up by the countervailing military successes of Alexander. It still exists, more than twenty-five hundred years after its founding, among the Parsees of India, a sect noted for industriousness and devoutness. By far the greatest influence of this nonhostile religion, however, came from the absorption of some of its concepts by Islam, Judaism, and Christianity. Those spooky Magi arriving camel-borne on the Christian Nativity scene are one small example: Gaspar, Melchior, and Balthasar were Zoroastrian priests. It is a curious characteristic of revealed history that matters can get twisted 180 degrees from the way they began. In the Middle Ages, Zoroaster was queerly revered as the founder of magic, astrology, and alchemy, which is about as ironical a rôle as could be imagined for a prophet who strove to lighten the burden of superstition on the human spirit. And at the end of the 19th Century, Nietzsche, who knew better, thought it perversely penetrating to portray Zoroaster as the first and greatest immoralist. On Nietzsche the tables have been turned more quickly than usual: the dramatic opening measures of Strauss's music for *Also Sprach Zarathustra* are almost all that remains in general memory of that dark work, and even they survive only as eerie background music for brave ventures in the conquest of space.

¶ CONSIDERING how much it has been stared at and pondered, fire has been singularly resistant to philosopher-scientists. It took thousands of years to develop a workable, self-consistent theory of just what fire was. Even today, more than a century after the attainment of this milestone, there are details—the fine structure within a candle flame, for example—that have not yet been fully resolved by chemists or physicists. If you are tactless enough to inquire about this, it can be predicted with confidence that a chemist or physicist will say that he has *much* more important matters to investigate than a candle flame.

One of the first professional fire-starers, and one of the best, was Heraclitus, a morose Greek of the 5th Century B.C. Almost nothing is known directly about him, our information being limited to the recollection and quotation of others; and it is wonderful that so vivid a picture of a human being can be transmitted to us by multiple reflection across a chasm of time. He had an unconcealed distaste for most other philosophers: "The learning of many things teacheth not understanding, else it would have taught Hesiod and Pythagoras, and again Xenophanes and Hekatios." He had a taste for gnomic observations: "Souls smell in Hades . . . every beast is driven to pasture with blows . . . asses would rather have straw than gold . . . oxen are happy when they find bitter vetches to eat." It is conceivable that he could have been influenced by Zoroaster when he concluded that fire was primal, the essence of change, and it was the only constant, unchanging thing in the otherwise wholly inconstant universe. This could, if you like, have been no more than one more gnomic remark by a grumpy seer at Ephesus. But it takes only an elementary translation of terms to change this into a remarkable pre-perception of the law of the conservation of energy, the cornerstone of modern physics.

Later natural philosophers, in the Orient and Meso-America as well as Greece, preferred a theory that included earth, air, and water with fire as the quadumvirate of basic elements. This is a formulation of no great discernment, but it was glib and ambiguous enough—the two desiderata of received wisdom—to survive for more than a thousand years. The next time a theoretician is observed putting on airs, taking a lofty view of grubby engineering as opposed to the intellectualism of pure science, it is instructive to think of the span of time when men deftly used fire to create glass and ceramics, to smelt, alloy, forge, and temper metals, all unsupported by any better theoretical underpinning than this formulation.

It wasn't until the beginning of the 18th Century that a Dutch physician named Boerhaave made an important observation. He wrote that the two main characteristics of fire—light and heat—are not precisely coexistent, for as a fire goes out its light ceases, but some heat persists for a time. Heat was the essential characteristic to be understood. By the end of that century, after Priestly discovered oxygen in 1774, the exothermic chemical reaction of fire was beginning to come clear. Once again theoreticians took a wrong turn when they postulated that heat must be caused by the presence of an invisible weightless fluid called *caloric* (the word is a noun, not an adjective). The theory of caloric conveniently solved some minor problems, such as the expansion of most materials with an increase of temperature. But like the concept of ether that was later to bedevil astronomers and physicists, caloric created vastly more problems than it solved. By 1798 and 1799, working separately, Count Rumford and Humphry Davy carried out experiments indicating strongly that caloric could not exist. Thanks in part to the momentum of sanctioned hypothesis, however, it took another forty or fifty years for the concept of caloric to receive formal burial. It was replaced by the concept of heat as the energetic movement of molecular and atomic particles. By mid-19th Century both experimentalists and theoreticians agreed that heat was a form of energy, and energy, if accurately measured and traced, was universally indestructible. Neither Zoroaster nor Heraclitus was referenced.

¶ SO FAR AS is presently known, everyday, run-of-the-hearth fire is peculiar to the Earth. No campfires light the dark of the Moon, and no moonwolves distantly circle the flames, stopping to bay at our white-whorled blue planet as it rises above a blocky crater rim. It can be stated with high confidence that there is not one single fireplace anywhere on Mars. Nor on Mercury, Venus, Jupiter, Saturn, Uranus, Neptune, nor Pluto. Nor in all probability* on any of the thirty-two known natural satellites in the solar system, nor on the thousands of asteroids that swarm about the Sun, nor on any of the strange comets that slope in from the unknown to pay parabolic obeisance to our star.

If you interrupt to say that the Sun is fire itself, well, that's really not quite right. By present scientific under-

* We shouldn't be flatly cocksure, not just because it is bad luck but also because our spacefaring achievements to date are about equivalent to the seafaring deeds of those ancestors who first paddled out astride a log to the nearest offshore island.

standing the Sun is a gigantic thermonuclear reactor, 864,000 miles in diameter, with interior temperatures in excess of 22,000,000 degrees Fahrenheit. It is simply not a large ball of domestic fire. Its commotions, far from consisting of heat and light, include prodigious amounts of every known kind of electromagnetic radiation, including X-rays. It is infinitely more protean and formidable than the friendly little self-sustaining oxidizing reaction that we sit around of a winter's evening.

Fire on Earth is possible because of the composition and density of the Earth's atmosphere. Where there is no atmosphere at all, or where it is composed of inimical gases, Earthly fire is not possible. In fact it is not easy to light a fire at high altitudes on Earth. In La Paz, which is above 13,000 feet, civic pride called for the purchase of bright fire engines, but at that altitude the partial pressure of oxygen barely supports fire and the chief problem is to keep the esteemed machines from growing cobwebbed and dusty in their firehouses.

Of course it depends on what you mean by fire. Volcanoes exist elsewhere than on Earth—Mars has a granddaddy of a volcano that dwarfs our biggest—and we use fire to propel our space ventures, carrying along the oxidizer necessary for burning. But in the solar system we apparently will not encounter the familiar fire of our home planet. On Venus and Mars the principal element in the atmosphere appears to be a gas that is used on Earth in fire extinguishers. This can pose no problem on uninhabitably hot Venus (so hot that tin and zinc, if present, run like water); but on antarctican Mars, the only fires that men from the planet Earth will hold out their gloved hands to will be imaginary ones, existing in their memories.

The Earth-limitedness of fire is for me an uncomfortable idea. Fire seems so profoundly a part of human life that it is unsettling to recognize it as a local phenomenon, a chance reaction made possible by a characteristic of one of the gases we paddle around in. Now that we have broken out of the atmosphere, the solar system is going to be the new neighborhood of man, and in its firelessness and other traits it will clearly be a strange place. Of course our fire could exist on other planets circling other stars. But entirely different reactions might exist there instead, and you are at liberty to imagine green fires that smell of methyl, that make a high squeaky sound, and that *absorb* heat. If they exist, such bizarre reactions hardly merit the sacred name of fire. Earth chauvinism is obviously going to be a problem in space.

The matter is not altogether trivial. Serious scientific thought has been devoted in recent years to the question of how man can communicate with extraterrestrial intelligence. This may not rank high on most people's list of urgent problems, but its seeming triviality diminishes when you hear scientists—solid, establishment men, not flat-Earthers—unfold the current estimates of probabilities. They construct a persuasive chain. Of all the stars in our galaxy, *n* have planets. Of these, *n* planets orbit in a habitable ecosphere, neither too close nor too far from their star. Of the number properly located, *n* planets have existed long enough for, first, the development of structured forms of replicating organic molecules called life to appear, and, second, for life to evolve into intelligent forms. Of these, *n* planets have intelligent life that has had the wisdom and good fortune to evolve advanced and stable civilizations. So enormous is the initial number in the chain known to be, that, even after conservatively large reductions are made at each intervening step, the sober estimate remains that there may be a million surviving advanced civilizations in our galaxy alone.*

It is not reasonable, most scientists aver, to expect to visit or be visited by beings from advanced civilizations: the immense distances impose impossible time and energy demands. But communication is not impossible, very likely by giant radiotelescopes. Some men imagine that the Others could well be drumming their fingers (if they have fingers), waiting for men to grow up enough to join the cosmic party line. The linguistic difficulties of creating a meeting-ground of meaning would not be insuperable. We would begin by using mathematics as a code key, on the assumption that mathematics operates identically on the planets of, say, Tau Ceti or Epsilon Eridani as it does on Earth.

Each time I have heard lecturers unfold this scenario, I have trouble with what are really small details. It is

* On theoretical grounds, advanced civilizations have already been divided into three categories, according to the degree of environmental mastery achieved. TYPE I civilizations control their planet. TYPE II civilizations control their star. TYPE III civilizations control their galaxy. The theoreticians who have conceived this mind-boggler are Russian.

not so much the diplomatic challenge of making small talk with a superior intelligence of totally unknown form and attitude. Nor the problem of chatting in Binary, which I suppose we could become accustomed to. Nor even the extraordinarily inconvenient delay time, which may impose centuries between question and response. It is simply this: When we get down to a concentrated exchange of information, how can we possibly explain to the Others the full load of meaning that fire has for us? Can we bring ourselves to transmit that it is deeply comforting to Earthers to sit about and stare in a light trance at an oxidizing reaction of woody fibers? Can we expect them, for all their intelligence and experience, to avoid concluding that they've got some distinctly creepy creatures on the line and let's just hang up and swing the beam to the next promising planet on the list? This could be a tragedy for man, an unappealable rejection by the intelligent cosmic community. And yet surely it should be risked, because an understanding of the intricately braided relationship between man and fire is a key step in understanding the species.

❡ CERTAIN ASPECTS of fireplace fire offer no mystery at all. Most animals are heat-seekers, like the cat dozing on a radiator cover or the dog flopped down on a sunlit patch of rug. Nothing is wholly simple of course, and the dog may additionally like the sunny spot because it invites human attention, and the cat will choose the radiator cover situated by a window that permits neighborhood surveillance, since cats detest being uninformed. But if the day turns cold and a cheerful fireplace fire is kindled, almost any dog or cat in a relaxed mood will immediately come up and lie by it. A Lhasa of my acquaintance, an agreeable creature but no mental giant, has always had to be restrained from curling up in the fire. No doubt some heat-seeking is instinctive, a lazy comfortable way of economizing on metabolic energy. Often domestic animals, perhaps misled by the thermal lag of a fur coat, go into a cycle of curling up too close to the fire, growing too hot and retreating to a cool corner of the room, and then repeating the process. But we don't behave so differently, except perhaps for the repetition, when we come stomping into the house on a bitter day, wasting no moment to escape the crying of the winter wind.

The feast to the eye is the second part of the lure of a fireplace that offers no difficulty of understanding. Those who have been blind from birth are unlikely to care strongly about open fires, and no wonder, because fires seem to have been created to be stared at. One character of Truffaut's attributes the popularity of television to the fact that "people seem to want moving images after dinner" and not enough homes have fireplaces any more. The blackened fireback and framing of a fireplace is a perfect small theater for the light show. The tempo of the flamedance is equally perfect; imagine what a tedium a fire could be if flames of uniform color moved only imperceptibly, like the hour hand of a clock, or if events within a fire took place in an unexaminable blur of speed. On its domestic stage a fire puts on a show of great sophistication, for the main outlines of the plot are comfortably familiar but the dialogue is always fresh and new. I have been attending this theater for more than half a century and have never once seen the same performance. Visually a fire shares with composers like Bach and Haydn the intellectually satisfying device of richly embroidered variations on simple themes. This is a visual device that Nature is fond of; one thinks of the varied linear motions within a waterfall, a willow tree dancing in light airs, the rip of breakers coming diagonally to a long beach, and the spumy assaults of an ocean swell on a rocky coast. But fire has still another appeal to the eye. These are the incandescent forms and structures at the heart of an established blaze that are so seductive to our imaginations, luminous caves and shimmering temples, a glowing Rorschach challenge we can rarely resist studying.

A fire-starer's trance also presents no puzzle, at least on a surface level, because the radiance of fire supplies a direct explanation. To the sense of somnolent comfort induced by infrared radiation, and the incensey smell of woodsmoke, and the soft fluttering murmur of flames, add a fixation on flickering light: this is an excellent prescription for inducing an hypnotic state. Leigh Hunt, who was surely in a position to know, declared that fire is a great opiate. The problem about a fire-starer's trance is not so much why it occurs as why it normally induces a spell so light and easily snapped. There is no difficulty in dozing by a fire, but it doesn't often lead to the deep, hours-long sleep, the kind from which we return as from a far country. When there is a fireplace in the bedroom, or when on a camping trip the sleeping bag is near the fire, the commonest behavior pattern is to arouse at moderate intervals, perhaps simply to check on or replenish the fire, and perhaps because we are instinctively wary about giving way to profound unconsciousness in the presence of fire.

Certainly an important element in fireplace fire is the breadth of our associations with it, both individual and cultural. Beyond question I am much influenced by memories of the bright fireplaces in my childhood house, which fill a storehouse of recollection. They begin at what must have been an early, crawling stage with the brass-polish smell of fireplace fenders and the employment of the big firewood basket by one hearth as, variously, a chariot, a cave, and a castle. There was a prohibited but joyous floofing of ashes with bellows. There was a rite of being brought indoors from blue winter dusk, summoned from engrossed tunneling in snowbanks and being placed like a small snowman on newspapers spread out before a great blaze in the front-hall fireplace, numb, half frozen, potentially cranky. (We wore woolen clothes then, on which an alligatored sheath of snow-ice promptly affixed itself, and an aunt or mother coped with the snow-covered buttons and ice-encrusted overshoe clasps.) I remember standing before the blaze, sipping almost undrinkably hot cocoa and allowing lotion to be put on my cheeks, burning from wind, cold, and youth. There was one evening before Christmas when I emotionally insisted that at least one fireplace and chimney *had* to be kept dark,* and later there was the time that Christina set fire to the dining-room chimney by overenthusiastic burning of Christmas garlands. (We all ran about shouting at each other, but no harm was done beyond a slight darkening of the paneling below the mantel, and some decrease in Christina's prestige outside the kitchen.) There was the day my elder brother excluded me from wood-splitting with the remark "This is a man's work," an observation he shortly underscored by splitting his toe, and needing as a consequence to be rushed off, slate-faced, to Dr. Pratt. There were uncounted hours of reading on the floor by the fireplace in the study, not from Lincolnesque dependence on firelight so much as because it was comfortable to lie by the fender, poring over such favorites as the page in the encyclopedia that showed the Flags of All Nations, or the cutaway drawing of an *Unterseeboot*, or the translation of the Iliad with an exceptionally graphic drawing of the Trojan Horse. I suppose this is why Hector and Achilles have always been inextricably associated with woodsmoke.

Other childhoods may have had less firelight in them, but few of us wholly escaped memories of fire at a time when our senses had the special acuity of youth. Skating always meant a fire on shore to thaw fingers and toes gone numb. When we skated the seven miles along the

* It seems likely that I was not so much concerned about the risk that the old gentleman's whiskers might take fire, or even that damage might occur to the green Lionel train, as the hideous chance that, seeing all chimneys trailing smoke, he simply might not stop at all.

frozen river to the dam, a round trip that could barely be achieved before dark on a winter afternoon, there always were riverbank fires every mile or so where we could stop for warmth and shy sociability. When the wind permitted, we opened our mackinaws and sail-skated excitingly through the dusk, a procedure that made warming stops particularly welcome. All children who had an Indian interlude, perhaps with Ernest Thompson Seton as trusted guide, will have memories of campfires constructed according to careful prescription, on which perch or sunfish were ritually cooked and ritually pronounced delicious. At summer camps across the land there were evening bonfires with choral efforts at "Clementine" and "Keep the Home Fires Burning." And when we were a little older those of us with access to a driftwood-strewn beach soon discovered the exhilarating courtship explorations that were possible around a leaping beachfire. Taken in all, it is impossible to count the number of ways in which we have been culturally conditioned to think well of a blazing fire.

Some lines from *Snow-Bound* capture the attitude exactly. In 1866 Whittier was writing in age about a vivid recollection of youth, of an episode that had been exciting and a little scary at the time, and then almost heroic in his memory:

> As night drew on, and, from the crest
> Of wooded knolls that ridged the west,
> The sun, a snow-blown traveller, sank
> From sight beneath the smothering bank,

We piled, with care, our nightly stack
Of wood against the chimney-back,—
The oaken log, green, huge, and thick,
And on its top the stout back-stick;
The knotty forestick laid apart,
And filled between with curious art
The ragged brush; then, hovering near,
We watched the first red blaze appear,
Heard the sharp crackle, caught the gleam
On whitewashed wall and sagging beam,
Until the old, rude-furnished room
Burst, flower-like, into rosy bloom;
While radiant with a mimic flame
Outside the sparkling drift became,
And through the bare boughed lilac-tree
Our own warm hearth seemed blazing free.

The pleasing aspects of fireplace fire considered so far, its warmth and visual rewards, its gentle trance and felicitous associations, must be coupled with other associations that are less easy to comprehend. Why, for example, does the fire-trance so frequently loosen our normally firm grip on the present? Staring at a fire often sets a mind drifting free in time, sometimes forward and sometimes backward, as though flames were a drug that is specific for the temporal sense. Other methods of altering consciousness do not as a rule modify our sense of occupying an instant of now, even though they have the power to make it a very queer now indeed. Yet it takes only a few moments before a hearth to demonstrate that there is an unpredictable time machine concealed in every fire. It might be expected that in front of the leaping flames the young dream chiefly of the future and the old only of the past; but I cannot confirm from observation and inquiry that any such neat classification takes place. In a fire-trance we are simply loose in time.

Another strangeness is the way men the world over have independently chosen to make fire a symbol for life. On an earlier page the ubiquity of the fire/life metaphor in language was noted, but the concept is more than simply linguistic. Zoroaster lighted a fire as a symbol of immortality, specifying that it was to burn eternally; and it burns in cenotaphs and memorials all over the globe, attended by people who never heard of Zoroaster. The symbolism itself—neat, holy, and readily comprehended—offers no difficulty. What is surprising is the fact that it is virtually universal among men, a species not otherwise remarked for unanimity of opinion, and the fact that it has won out completely over such other symbol-stuff in nature as dawn, spring, tide, and river current. We even employ a kind of temporary eternal fire to denote the periodic existence of an institution—not even a living thing—in the fire ritual that marks the beginning and end of Olympic Games. Here of course the custom may have found root not so much in anything as formidable as immortality, as in the flaring pagan theatrics that Hitler found so suitable in his massed prewar rallies.

A dark theme woven into man's relationship with fire is something similar to but not identical with fear. We do not forget that the friendliest little fire can burn us, and grow to be an uncontrollable monster. Twelve thousand Americans die accidentally by fire every year, and the gigantic firestorms of war have been killers of unexampled scale. Yet the fundamental feeling is not purely fear, because if it were we would never let fire out of its cage. As shown when we admonish children not to play with matches, our attitude is a complexly structured one that recognizes risk while it admits attraction—an equation that, in balancing peril and pleasure, resolves to a kind of submerged wariness. Perhaps it is this balanced tension that works to inhibit the deepest forms of sleep in the presence of fire. For all men, risk is also pleasure, and one remembers poor Louis Slotin playing with his pieces of subcritical plutonium, twisting the tail of the dragon that breathed more than fire. Psychiatrists have theorized that some measure of playing with fire is a factor in the behavior of habituated smokers, a notion that, as you watch a pipe-smoker clean, load, and fire a favorite pipe, puffing like a locomotive and tamping the glowing coal, has less implausibility than many ideas vended by the practitioners.

Another aspect of fire, suited for consideration some blowy night when birch and maple are glowing on the andirons with radiance stored from summers past, is why fire came to be linked in men's quirky minds with the supernatural. In a way, the powerful and rationally behaving domestic god that defeats darkness and frightens predators might have been thought to be antimagical. If by some chance fire had been domesticated only in the last few thousand years, its associations might have gone differently, as implied by our ready acceptance of the Graeco-Roman use of a lighted torch as a symbol of learning and reason. But fire came to man's hand more than a hundred times a few thousand years ago. At some far distant time fire became linked with the spirits and demons that harried and haunted man. The event is entirely unknowable; all we now can do is to piece information with intuition. One piece is contributed by cultural anthropologists who note that obsession with fire, often observed among primitive peoples and psychopaths, has characteristics that suggest it may be of very ancient origin. There is also indication that in ancient days fire was used as a stimulant to arouse and excite, in contrast to our use of fire as a comforting sedative. Finally, many deep and spacious caves occupied in prehistoric times were seemingly not used for everyday shelter, for they could be reached only by crawling through long and difficult passageways. They seem to have been special places set aside for unknown ritual activity. We are unlikely ever to know the pre-barbaric torchlit scenes that took place in these smoky caverns, where fire was perhaps the powerful drug that led our remote ancestors into rituals of collective excitement.

The last characteristic of fireplace fire that must be mentioned is one of such tenuous subtlety that it may not even exist at all, and certainly not for everyone. But when I study a well-behaved fire and submit without resistance to its spell, slipping temporal moorings and drifting, not backward in time to barely imaginable rites in a smoky cave but instead forward, I think it is possible to see in the shimmery cubical structures at the heart of the blaze that we are very far from through with fire.

The time scale of this perception—perhaps focal length is a better term—is far past transient local perturbations, far beyond energy shortages and the technologies of central heating. It extends into more remote regimes when other events, perhaps the decisive end of the current Interglacial episode, occupy the concerns of men. Although the details are unclear, it is plain to me that we are not yet done with fire.

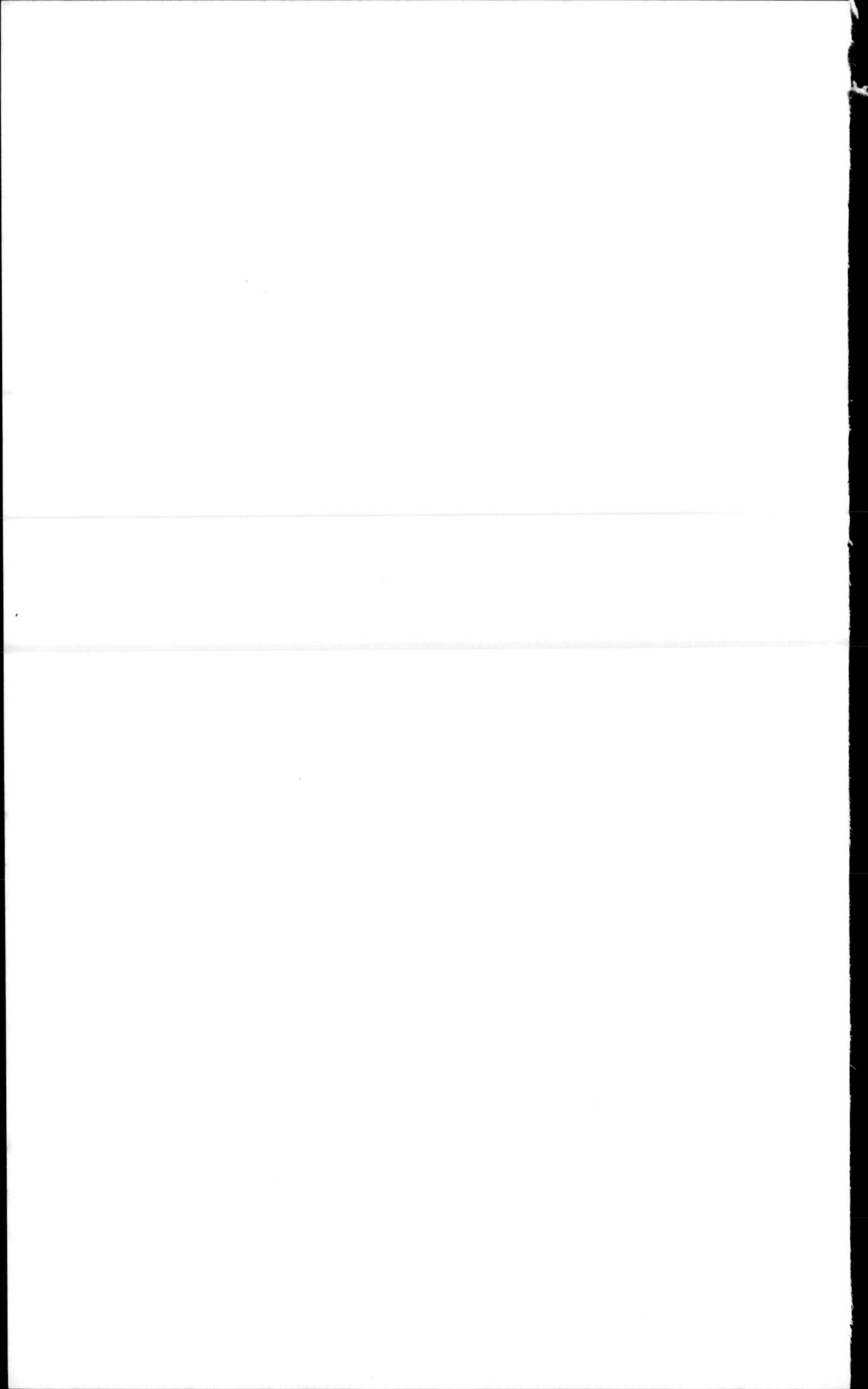

AFTERWORDS

THIS book was begun years ago, when energy was a run-of-the-mine noun, and profuse amounts of automatic central heating were part of that curious grail called the Good Life. Its purpose was simply to share some of the fascination that fireplace fire has always had for me. Thanks to the slow tempo of after-hours writing, it emerges into a different world; but fire is no less enchanting for being useful.

Readers wishing to explore byways will find that John E. Pfeiffer's *The Emergence of Man* is a first-rate introduction to prehistory. An Englishman, Walter Shepherd, has recently published *Flint*, a stunningly comprehensive book (Faber). Vrest Orton's idiosyncratic *The Forgotten Art of Building a Good Fireplace* gives the proportions of Rumford's fireplaces; and many rich details about that unlikely man can be found in W. J. Sparrow's *Knight of the White Eagle*, published in this country as *Count Rumford of Woburn, Mass.* Of the hundreds of books about Franklin, Carl Van Doren's biography seems among the most discerning and rewarding.

For this volume, nuggets of information and steadying counsel have been contributed by Jean Allaway, Mary Anderson, Doris and Roland Coryell, F. George Drobka, Janet and Stephen Greene, Wilhelmina Kallenbach, Oran Nicks, Fay and Frank Rowsome III, Volta Torrey, Peppino Vlannes, Kay Voglewede, and Van A. Wente. Years ago Hugh Coryell and my father devoted much patience to sharing with me their woodlot and fireplace skills. Any errors of fact or eccentricities of opinion are, however, entirely mine.—Frank Rowsome, Jr.